The Colliers Tattletale

By:

Ronald H. McIntyre

Published & Distributed

by:

FORMAC PUBLISHING CO. LIMITED

Box 1688

Antigonish, Nova Scotia

TUESDAY, JULY 20, 1982
ON TRIP FROM KELTIC LODGE IN INGONISH THRU BADDECK TO PORT HAWKESBURY.

FIRST EDITION, March 1980

Published by:
FORMAC PUBLISHING COMPANY LIMITED
Box 1688
Antigonish, Nova Scotia
Canada

ISBN # 0-88780-003-3

ACKNOWLEDGEMENT

This publication was made possible with assistance from the Nova Scotia Department of Culture, Recreation and Fitness.

Contents

DEDICATION

This book is dedicated to my brother-in-law Thomas J. Casey. "Tommy" was one of the best-loved persons in the colliery towns of Dominion, New Waterford and Glace Bay during the years he was a mine manager in Cape Breton. Later, Mr. Casey was appointed the Chief Inspector of Mines for the Province of Nova Scotia.

— Ronald H. McIntyre

Introduction

Life was difficult and often cruel in the early coal districts, nonetheless, mining towns have withstood its shortcoming and many residents have advanced their education in colliery mining schools since the first of this century. I finished grade school at St. Anthony's in Passchendaele, Glace Bay, and one term in a private school in Cheticamp, Cape Breton. Most of my 'reading, 'riting and 'rithmetic was advanced in night schools. During my adult school days I took several courses in "Job Training", and Industrial Labor Relations. In later years I attended the Ford Industrial School, in Detroit, Michigan, for one term. It is my thought throughout this writing to encourage colliery residents to promote mining classes in their own towns for their advancement in safer mining methods. This small book may be a guide for miners whose fundamental knowledge of mining is much greater than mine, because too often people forget the significance of underground "housekeeping", which reduces accidents and increases coal productions.

During my research of this book, people in public archives and libraries assisted me on many late Saturday afternoons. I earnestly thank them. I am grateful to the people who checked my writing and encouraged me, and I have a very special thanks for the people who worked with me in the coal mines. The pictorial history of underground mining has involved the final research for this book, and I was very fortunate for the kind help from the people at the Glace Bay Miner's Museum — thanks fellows for the photo copies. Photographs will unlock history and display our heritage, and it is possible in the near future to have a "Pictorial Mining Story".

Ronald H. McIntyre

The Miner

The men that sail the ocean
Know the perils of the sea,
Yet the man that is a sailor
A sailor will ever be.
Lured by the whispering billows,
How can they understand
The heart of the man that toils below
With a shovel in his hand?

Trainmen love the chugging engines
Speeding o'er the shining rails,
While the hunter in the forest
Walks alone in silent trails.
While the brawny muscled farmer
Guides the ploughshare o'er the land,
Down below the black-faced miner
Swings a shovel in his hand.

Deep within old Nature's bosom
She hath stored the precious coal
Reaching out beneath the ocean
Where the mighty billows roll.
Far beneath the city dwellings
And the peaceful country land,
Stands a lad to fill the boxes
With a shovel in his hand.

Up above the sun is shining,
In the mine it's always night;
Like a firefly in the darkness
He can see his buddy's light.
In the front line of the battle
He may never take a stand;
He will strike a blow for freedom
With a shovel in his hand.

He will keep the home fires burning,
He will proudly do his bit,
While you pay in easy money
He will pay in toil and sweat.
Dangers lurk amid the tunnels,
Many times the miner stands
Looking at the roof above him,
With a shovel in his hand.

With his lunchbox on his shoulder
God's blue sky looks good to him
When he comes up to the surface
From the shadows dark and dim.
Laughing, joking in the washhouse,
Scrubbing palms with soap and sand,
For the man that wields the shovel
Bears the imprint in his hand.

He will grumble at the check-off
While he plans to give some more,
You will find him liberal-handed
When you tally up the score.
Don't forget the Glace Bay miner
Is the finest in the land —
He's the mainstay of Cape Breton
With a shovel in his hand.

—Lillian Crewe Walsh

Preface

This book is an attempt, by a former colliery resident and coal miner, to explain the mining knowledge acquired from home study and the practical mining schools in Atlantic Canada. Coal will become once more a major industry in all parts of the world, and it is essential that young men from the colliery towns who have attended night school mining classes for three or four years have the supervisory positions in the mines. People with mining experience in the pits and have passed two or three of the written exams which is necessary to hold management jobs at the Nova Scotia collieries will reduce accidents underground and increase corporation profits. Unprofitable coal companies have no security for their employees. So, much of what I write in this book will be a story of deep mining operations, "and the ever increasing need for mine supervisors who have gained their mining education in the colliery mining schools."

This writing represents several years of study and research by the author, and the writer has experienced much of the material written into this book. It contains fundamental information necessary to carry on knowledgeable mining operation in gassy coal seams. It is long known in mining circles that people schooled in the laborious part of extracting minerals from the earth are coal companies best supervisors. However, it may be expedient for executives from outside the mining districts to hold dominant positions in the future to solve problems when wages and profits curtail the safety of mine workers. However, the expeditious efficiency of aggressive businessmen could have an adverse affect in mining communities.

It is possible for large corporations to find management among the working people after they have established industrial schools for their employees.

The problems which will confront the future mines and the mining communities will be many, and an organized system of colliery mining schools for the colliers is necessary in most of the coal districts in North America. Although the Province of

Shortage of managers is forecast

TORONTO (CP) — A new study reveals that Canada is faced with a shortage of managers that could lead to economic disaster in the 1980s.

The study, commissioned by the Social Sciences and Humanities Research Council, was made public this week at a conference sponsored by the Canadian Federation of Deans and Management and Administrative Studies.

"It is becoming increasingly evident that because of the state of management education, the Canadian economy is entering the 1980s at a severe disadvantage," the study says. "The condition of management education has reached crisis proportions."

It said without adequate funds and larger staffs, management schools will not be able to graduate enough professionals to meet future demands of industry and government.

Nova Scotia has supported the mining classes in their coal towns for over fifty years, this writer was told during the Winter of 1978 that only a few miners were in the schools that year, and the attendance would soon be zero. The vital question now is, are the mining classes in the town of Glace Bay, the center of Coal Mining in Cape Breton Island, going to be abandoned by the lack of new students? It would seem that the attentions of young colliers in the mining districts will allow themselves to labor in gassy mine workings without any other knowledge of mining safely than which is learned from experience. The orgin and the actual workings of these types of schools in mining towns has an economic value in the communities in which they exist, and they should not be allowed to pass away in Nova Scotia coal mining districts. No better advice can be given a young miner — "attend the colliery mining classes."

The mining communities need people from the mines with ambition to advance in the coal industry and the union. The miners should take a short course at the mine schools even though he may never want a job as a mine boss. The new mining student is generally an adult with a fair amount of mathematics behind him. People who have not attended school for many years should "brush up" their arithmetic before registering at the mining schools.

REGISTRATION
Coal Mining School
1970-1971

Operated by the Department of Mines
Province of Nova Scotia

Will Be Held

FRIDAY and SATURDAY

SEPT. 18 and 19

TIME: 7 P.M.

AT

Glace Bay Junior High School
MacLean Street

AND

Donkin School, Donkin

CLASSES AS FOLLOWS
1-2-3 Class Stationary Engineer
1-2-3 Class Engine Operator
Coal Mining, Mine Electrical, Mine Survey
For further information, contact local

WANTED
COAL MINERS
At The New Coal Mines
McIntyre Porcupine Mines Ltd.
Grand Cache—Western Alberta

Representatives from the Company are now in Glace Bay to meet with interested men and wives.

Contact can be made during the month of Nov. - Monday to Friday, 10 a.m. - 5 p.m.

CANADA MANPOWER CENTRE
Department of Manpower and Immigration

59 MAIN ST. 849-3541 GLACE BAY
(Post Office Building)

Province of Nova Scotia
Department of Mines

MINING EXAMINATIONS

Examinations for Certificate of Competency as First Class Mine Official (Manager); Second Class Mine Official (Underground Manager); and Mine Surveyor will be held on:

MAY 26-27-28, 1977

Candidates for Examination must make application in writing to Secretary of the Board of Examiners, Dept. of Mines, P.O. Box 1087, Halifax, Nova Scotia, B3J 2X1, not later than May 16, 1977.

Candidates will be notified of the place and time of writing.

A. ARTHUR MACDONALD
Secretary
Board of Examiners

BN-344

MPs want action on coal stepped up

Coal mine accidents cost $57 million

WASHINGTON (UPI) – Coal mine accidents cost workers, companies and taxpayers nearly $57 million in 1974, with the biggest share of the cost borne by injured miners and their families, the Interior Department said Sunday.

was gathered by visits to coal mines, insurance companies, mine inspection offices, workmen's compensation agencies and medical centers.

The writer learned this from his own experience in the schools. The extent of the students' ambition and their abilities to come to grips with their books at home, and the mining teacher's competence to teach will be an advantage to miners who work in the pits and in their communities. Maybe some reform is needed in the Nova Scotia mining classes in the coal towns, and one of the disadvantages was the division of students in the schools. Although the classes were divided into two groups, new and older students, the teacher was trying to instruct both at the same time. New students were listening to curriculums taught older miners; this has discouraged young colliers and many became "drop-outs" over the years.

However, we are often told the educational treatment for a mining study is primarily one for universities. That is common for many of the mine managers in the United Kingdom after they have acquired mine certificates from their teaching in the colliery night school classes. No person shall have charge of any coal mine in Nova Scotia unless he has a mine manager's certificate that was duly authorized by the provincial mining department. Some of the students that learned about mining in the colliery mining schools became mining engineers, but most of the miners who became mine managers in Cape Breton never had time away from their jobs to attend the technical college in Halifax. The Nova Scotia colliery managers have advanced mining methods and safety in their coal pits. The Cape Breton coal mines had at one time the best safety record in North America. The annual mining examinations are held in Nova Scotia each Spring for the colliers who are trying to advance themselves in the mining game, and for the safety and security for their own people and their livelihoods.

Mining certificates should not be given to people who are appointed underground supervisors by oral examinations, although it is a habit in mining communities that do not have mining schools for their miners. Oral examinations of people for mine "fire-bosses", who never made a study of mine text books or attended a colliery mining class, is not enough preparation for an underground supervisor when an emergency occurs in his section of the mine workings. The section boss must quickly recognize the circumstances of gas accumulation or fire, and have the ability and courage to handle unforeseen occurrences before they develop into major incidents. The inexperienced mine boss has no immediate help from the surface people when communication into a "blocked-off" mine workings are "blacked-out". The underground supervisor who is charged with the safety of people during the hours which coal is removed from the faces should be capable of writing and passing a mine overman's exam which is required by the British Mines Act. The English Mines Act is generally a basic for other countries' mine safety acts.

Coal is taken from the ground by large investors for commercial trade; industrialists and governments supply the capital needed to open and operate mines. Before the 1500s in England, workmen dug coal from shallow coal seams for their own use and some had a small coal trade in the villages. Later the coal leases were given to wealthy barons, and miners were working for a few pence a day in gassy underground atmospheres with open flame torches.

It is not the intention of this writer to be so critical of coal companies to make them look like damnable corporations which used people for profits and then cast them aside like used materials. Nonetheless, certain mechanical differences at the collieries gave rise to this widely circulated rumor. Neither is it the intent of the writer to claim the miners were defenceless people, cowed by years of injustice, and incapable of striking back at unscrupulous mine operators of the past decade. Miners are not looking for something for nothing during strikes at the collieries, but they will demand a fair day's pay for a day's work.

The inaccuracies of press releases have falsified conditions in the mining communities and in the mines for many years, but there is reason to believe the colliers' prestige has extended beyond the coal fields. Already it is known that the "big husky miner is a gentle guy", and visitors into the mining towns have told about the cultural orderliness of the inhabitants. However, there are always a few people who like to dramatize on the worst of coal mining.

Because coal companies were concerned with their productions and costs at the "pit-heads", and often forgot it was people who created profits and losses, and by habitually regarding the laborer as an instrument of production, objections to corporations' principles were challenged in the mining districts of North America many times. It took years to forget faults connected with industrial work stoppages, and it is fact that both union men and company officials had to leave certain districts for parts they played in coal strikes.

Nova Scotia had more than a share of tragedies during mine disasters, and their pit yards on the surface around mine shafts were scenes of tearful eyes as lonely people waited for the loved ones to come up from the fiery mines. Some were left down there for a long, long time. Many years after a blast in the old Foord mine in Pictou County, miners buried their grandfathers' skeletons that were found in the mine workings. Most of the large loss of people in mine disasters in Eastern Canada took place in the 1800s on the mainland of Nova Scotia. However, Springhill, one of the oldest coal mining towns in Canada, had two mine disasters during this century. Some years after the colliery explosion at Springhill during 1891, that killed 125 miners, families from that town were moving into Cape Breton.

The average collier reads and listens to more current news than most industrial people, and miners may be the most informed persons of labor history among the so-called "blue collar wage earners". Miners read everything from the bible to sporting magazines, but the colliers of years gone by read history books. I have listened to miners during lunch breaks in the pits thirty years ago discussing "The Fall of the Roman Empire", "King Henry's several wives", and the retreat of "Napoleon" (Bonaparte) from Russia during the Winter of 1812.

This little book narrates the collier's way of life that has changed during the last few decades in the colliery communities all over North America. The coal is mined and loaded underground nowadays with fast-moving machinery, powered by electricity inside the earth's crust. The hours of work are shortened, and it is seldom the coal face miners are at the "faces" for more than six hours. "Company picnics" are no longer a holiday treat for tired miners who worked from the morning daylight until the evening darkness for one dollar a day. Nonetheless, there are still a few company towns and houses in North America where the union is collecting dues from the workers. Some years ago a Cape Breton miner journeyed westward on the advice of a coal company agent that he was going to a progressive town with new homes and schools. That collier soon returned eastward after he was housed in a mobile shack which was used for a tool room during the construction of surface buildings.

The author started to write this article to inform miners and mine bosses of the information that he learned from the study of mine textbooks at home and the teachers in the Nova Scotia colliery mining schools when he was a young man. Why am I doing all this writing at my age? Mostly because of the nonchalant attitudes of governments, coal companies and the unions since the end of the second World War when mine laws were not enforced in the mines. The United States Department of the Interior told its Senate during the month of June 1977, that inadequate enforcement of mine laws may have contributed to 260 deaths in the American mines during five colliery explosions. In less than two weeks during the month of April 1978, nine miners were daily fatalities in the State of Virginia coal pits. Virginia is the fourth or fifth State in coal productions in the United States. Late in the Fall of 1977, five rescue men

"THE COMPANY PICNIC TRAIN"

This was a "Labor Day" outing at Mira Gut which was sponsored by the "P. W. A.". The new union had several organizers at the picnic, and there were a few fights.

MINER'S PICNIC TRAIN CROSSING MIRA RIVER BRIDGE – 1898

were rushed into a burning mine to rescue a trapped miner in West Germany; a few hours later all six were found dead by the second rescue team. Methane gas and coal dust were ignited in a mine at Glace Bay, N.S., in February 1979, and 12 young miners are dead. It is time for the colliery residents to start asking questions and have public investigations to find out what is happening to their people in the mines nowadays.

It is not enough to think or say the blast at 26 Colliery in Glace Bay will be a lesson for others and it should not happen again. There are many questions to be asked and answered for the coal mines in North America to be made safe for the mine workers, and contrary to the thoughts of most people, there are safe mines operating and investors are happy with the results of their investments. Although some people are trying to kill "the goose", which is now laying "golden eggs".

There are many questions to be asked and some of the answers are in this little book. They are nothing new — many of the old mine managers in Cape Breton asked similar questions at the weekly safety meetings. Are the ventilating air currents underground obstructed through careless mining practices? Are miners operating electric mining machines on coal faces and known explosive methane is stuck in nooks and backs of the gob packs? Are some underground supervisors not capable of enforcing the mines act? Is it possible that miners in the Nova Scotia coal pits are given mine overmen's positions who do not own written certificates for a study they made on mining and supervisory methods? Are the mine rescue squads coached by intelligent people who supervise and check out the equipment worn by miners during under the ground fires? Two noted Nova Scotia mine rescue men were lost in Number 26 a while back through faulty equipment given them to fight a fire. Generally, when no explanations are found for serious accidents, it becomes an act of God, and most people go along with such answers. Are these questions of no importance to the general public or the colliery women to ask, "Is my husband, my son, my father, working in a safe mine?"

The first part of this book deals with the fiery rocks that people live upon approximately 25 miles below the surface atmosphere, and how it affects mining operations in deep mines. The second installment is coal miner's view of the earth's geology from a study of mining textbooks, and the teachers in the practical mining schools in Nova Scotia when this writer attended the mining classes in Glace Bay. However, the writer seized an opportunity here to explain the problems connected with removing coal from the hard rock strata in gassy coal seams. The third part is headlined "The Flame Safety Lamps". It is an interesting story of two men who claimed the invention of the lamps. Most of the writing on the lamps written in this book was never publicly known. Here once more the writer takes the liberty to write about the early days of mining in England, and the people who worked underground in near darkness for their livelihood. The last chapter of this little book tells of the problems in mining communities all over North America. The writer tells of his childhood days in Cape Breton mining districts known as "Passchendaele and Caledonia", and his father's farm on top of old mine workings which was or became mostly "pit holes". The company stores is a part of an old article which the author had published 25 years ago. Miners' sports, rugby and baseball, were played in Glace Bay before this writer was born. The true story of the horses which worked in the mine darkness without reins to control them, and the tiny young boys who became their "drivers" in the old Caledonia mine.

Ronald H. McIntyre

PART ONE

"The Crust of the Earth"

"GEOLOGY ILLUSTRATION"

CORES — — — MANTLE — — — CRUST

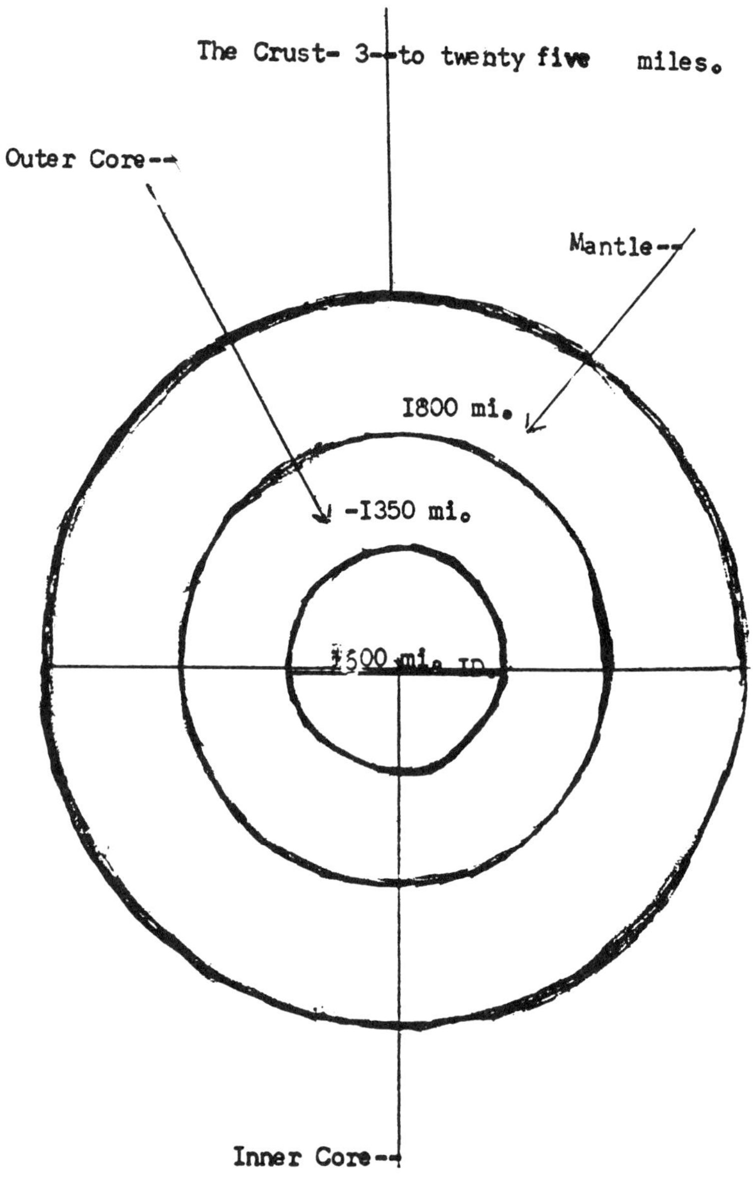

Part One

"The Crust of the Earth"

In order to describe the illustration sketched here by the writer it may be necessary to write several paragraphs to clarify its purpose in a mining article. It is not an essential study at the practical mining classes in the colliery towns. The explanation on what happened inside the earth more than a billion years ago is a much greater contemplation than this writer can tell or thoroughly rephrase. The breakdown will be brief, but the earth's mantle does have a certain connection with coal mining in "deep coal seams".

In spite of the fact that warm underground air becomes a discomfort to the miners working at the coal faces, air transmitting warmth in the mine workings does influence a movement of new air into mines for the workers to breathe. The temperature in the underground strata of a deep coal mine will become one degree warmer on the Fahrenheit scale for every 60 feet on a 90° angle the coal faces are worked towards the earth's mantle. A coal seam with approximately 700 feet of cover would be considered a deep mining project. However, heat from the hot mantle does not warm the surface soil. During the last century the natural air underground was moved about airways by increasing the temperature of the atmosphere returning out of the mines with fired furnaces. The bricked furnaces were generally placed at the bottom of "return air shafts".

The sketch pertaining to the inner and outer cores of the earth is considered by the ordinary person as the center of the conglomerated mass of rocks and minerals below the surface. Geologists have named the inner point of the inner core the "earth's central". The illustration should be carefully scanned by the readers to get a general idea of what is lying below the ground we live upon. It is an interesting tale and a little scary to learn that we are living upon a very hot fire, approximately 25 miles from the sufrace.

At one time scientists claimed the inner core of the earth was a molten mass, and the outer core about the same environment. Geologists now say that is not correct. It is now believed by most scientists the inner core is a round solid mass of hot materials, and the outer core a body of matter neither solid or liquid. The fact that different writers have adopted opposite views on the conditions inside the earth only serves to show the magnitude of this subject. The outer core which extends 1,350 miles towards the "central" is believed by some scientists to be more of a liquid than a solid. Scientists also tell us the earth's cores cannot become a molten mass because of the tremendous pressure locked inside the earth around its central.

The fire, which keeps our geological pot boiling inside the earth, does not fundamentally come from the hot cores, but rather from radioactive rocks in the mantle which are creating rivers of fire, not more than 25 miles below the ground where minerals are taken from the crust. The mantle had a depth of 1,800 miles, which surrounds the outer core boundry at a temperature of 4,000° Fahrenheit, and we are also reminded that melted rocks bursting from volcanic eruptions has a heat of 1,200° celcius. The earth's crust sits on that hot mantle, where movements are continually bombarding the lower boundary of the crust with shocks. Many of these shocks are not noticed on the earth's surface by its inhabitants, but scientists have the Richter scale to measure the intensity of earthquakes or light tremors when they happen deep inside the earth. Most geologists have the Geiger counter for indicating radioactive substances in the mantle lava after it is exploded from volcanoes.

Light tremors or shocks may precede an earthquake in regions where large faults have cracks in the underground strata which may train into the fiery mantle. Luckily most of these eruptions occur on the ocean floors. The most violent shock wave in North America was centered in the State of Missouri, during 1911, which rang church bells in far-away Boston, Massachusetts.

The crust is directly below the surface, and it is divided into three divisions of rocks for approximately 25 miles by geologists. The bottom 10 miles is mostly heavy black rock, and on top of that is 10 more miles of lighter stones and materials similar to the substances in the lower rocks. Then there are four or five miles of sedimentary stone which contains minerals. The earth is not perfectly round; it has several bulges from the equator to the poles, and there are deep canyons on the bottom of the oceans where the crust may have a much thinner rock base than 25 miles. There are many different kinds of rocks inside the earth. Some have great strength to withstand the most violent shocks, but there are rocks in the underground which are weak and broken by the materials in their compositions and the enormous weight the earth creates inside the rock strata. When we read or hear these words we feel we are dealing with people whose thoughts are not ours in a spiritual perception, but many of today's scientists were enlightened by the powers of the supernatural.

"A" Profile of the ocean floor between North America and Africa. The spreading and slipping takes at the edge of both continents and at the "mid - ocean ridge".

"B" This fault was produced and bared after the Montana earthquake, which happened on the 17th. of August 1959.

A

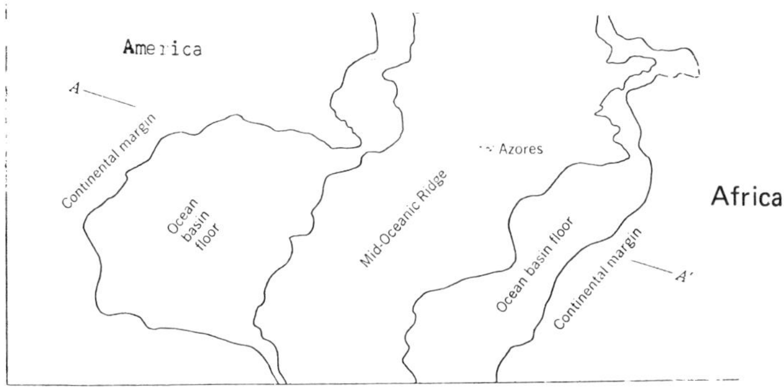

B

Twenty-three years ago, 1957, scientists from all parts of the world formed "The International Geological Year". The combined intelligence of the earth's foremost geochemists were going to gather information from the ocean floors by a mechanical thrust into the fiery mantle. It became known to these people there were several ocean floors where the crust was only five miles thick. Geological periodicals named the scientists' project "Mo-hole".

The Mo-hole project was scuttled before it really had a fair start, and its failure was blamed on the lack of trained people to work under the water, finance and the need of togetherness among the people who drafted the program. Scientists as well as others have learned that knowing how a job should be satisfactorily completed is not enough; workers must be taught and trained for special jobs under water and under the ground. However, geologists know more about the inside of the earth nowadays than ever before through the study of seismology. Nonetheless, they have never made an analysis of any materials taken from the lower boundary of the crust by mechanical operations.

Seismology isn't a theory on rocks. It is facts taken from an analysis of the lava which spurted over parts of the surface during volcanoes, and that rock is keeping scientists informed on what is happening inside the mantle. Through a study of seismology, and using an apparatus known as the seismograph, geologists can measure and record earthquakes by sounds which they create inside the earth. Soon after the shock wave is located on the seismograph, scientists gather there for information on the shock waves inside the earth. Most earthquakes do not expel hot lava over the surface, but they will destroy towns and cities during violent "shakes".

Directly above the focus of an earthquake on the surface the space is known as the "epicenter". The vibration from the wave point will travel into the earth in all directions. Scientists who made a life study of seismology have discovered that sounds artificially made at the epicenter on the surface will make a quick start, slow down and stop on its tour through the inside of the earth.

During unknown geologic eras, which are considered divisions of geological ages, the world we now live upon started to form a crust around the hot mantle. Throughout this building of surfaces, which took many millions of years, certain parts of the earth's stratum did not have the basic materials to build a solid base over the fiery mantle. That is one of the main causes which left the crust with so many faults and crevices. Eruptions in the mantle seek a way out, and they get relief through faults and volcanic pits in the surface soil. Most shock waves are centered on the ocean floors. There are high rock ridges in the Atlantic Ocean that geologists claim were caused by earthquakes many years ago, and most scientists think this has caused a spreading of the continents between North America and Africa.

Through great spans of time counted in millions of years, the world's continents were moving away from each other, and a few scientists think it is happening today between North America and Africa. The spread occurs in inches and feet over short periods of geological time, and people who live in the areas which have a weakened underground strata will notice small shocks over their land surface. However, the continuing spread of the underground strata, for years too numerous to mention, may cause large blocks of land to slip from a fault and cause a violent earthquake. It is now known that most faults in the earth were not caused by earthquakes; they formed in the construction of the strata. However, geologists have stated that known ocean ridges on the floor of the Atlantic Ocean are faults where earthquakes have occurred at one time.

Geochemists or seismologists who study the different sounds and vibrations created during earthquakes have stated the shock waves moved from the epicenter in opposite the curvature curve. The first wave which travels through the crust is the dominant pulse which often opens and splits the earth's surface. The second shock wave goes directly into the earth in all directions, but it is deflected 1,800 miles inside the earth. The third wave may be eliminated completely at 1,800 miles down. However, the second wave on another beam becomes active once more and continues through the 1,350 miles of the outer core at a slower pace, but gains momentum at the outer boundary of the inner core. The movement of sound through the earth tells scientists the nature of the materials in the mantle and the crust; one must be of a hard nature and the other soft. Sound moves quickly through solids.

It is possible that our earth moved through the heavens during unknown history of this world, bumped into other planets, scalded close to the sun, and froze in an icy era far from the sun's rays. However, power unknown to many people was building a world that became a mystery for the people now inhabiting the earth. The ball of rock below the earth's surface is composed of many elements of radioactive substances which contain "sun-dust". The sun dust has been smoldering and bursting into a fiery environment inside the mantle since the earth's creation. Scientists think it is acceptable to think that a great amount of the water in our oceans was produced from minerals in the top of the mantle.

The earth's crust as we know it nowadays has vast supplies of water and vegetation. There are 57 and one-half millions of square miles of land surfaces, but water covers more than 70 percent of the total area of the crust. Many of the great scientists believe that for many millions of years after the world was formed and divided into land and water no living thing inhabited this planet. Although the first environment and life are closely related in a study of geology, scientists nowadays are generally interested in the birth of life and how it happened and the first years of this planet's environment is less of a study for them.

There are many people who think nature performed the miracles that caused the creation of life and the world. No way could the natural forces of nature accomplish with such accuracy that gave life and this beautiful planet set as a rosette in a galaxy of glistening stars. The earth has many mysteries for its people to discover, and its greatest is the belief in occultism powers of the supernatural God that has created the world which supplies the needs of the people He created. There is a mysterious period in geological time that scientists have discovered — the construction of the earth was stopped for millions of years. During this great black-out it is possible the mystifying divisions between man and his creator is locked. Few things are more remarkable for thought than the creation of the world, and the periods which the earth's history is written theory.

Rocks containing microfossils of unknown primitive organisms were discovered, during 1977, on the sea coast of Africa. They may prove that life was a part of this earth's environment billions of years ago. Biologists have stated life was nonexistent in the world because there was no oxygen inside the atmosphere to support life, which was mostly carbon dioxide and nitrogen. It is also told that scientists have known for years that insects are living in warm dark places which could not survive in an environment containing oxygen. The people who are looking and cutting into rocks with powerful microscopes are telling us the ground we live upon is three and one-half billion years old, and the crust started to form around the mantle five billion years ago. Nonetheless, each passing decade, geologists are finding rocks containing fossils inside of them which are older than the ones analyzed at the turn of this century.

Most of the material inside the earth is rock, which was at one time a cloud of dust in an atmosphere of space, and scientists claim it was that dust which formed the base of the earth. Pure rock is considered by geologists a mineral sand, which does not contain vegetation or animal matter. Although there are many different kinds of stone in the ground, scientists have divided all rocks into three main groups, and named them igneous, sedimentary and metamorphic. All rocks of any nature are connected with those three in some way or other.

Igneous rock was formed from grains of sand at fiery temperatures, which cooled in the earth, and most of the granite rocks are igneous stones. Metamorphic rock may be any number of rocks which have been changed during great periods of time and eruptions were common in the hot mass which was moulding our earth. It also was heated and twisted into different forms of rocks, but most metamorphic stones are slate or marble. Sedimentary rock was formed under water, and it contains mostly sand, with various amounts of mud and other materials. The rocks which traverse the coal measures are sedimentary stones, some of which contain sea shells and the remains of sea animals. Coal is known to geologists as sedimentary rock, which confuses the ordinary every day mine worker who made a short study of mining geology. Rocks are mineral which has no plant life, but coal is a mineral which contains vegetation and animal matter. "Rock had a beginning as mineral, coal was mineralized in the earth."

The story of coal is theory which states that vegetation which grew on this earth during uncertain periods in geological history was pushed into the ground and later became coal after it was covered with sand and silt, and water for centuries. There are coal deposits in many part of the world, and several seams in each colliery district. The United States has the largest known coal beds in the world, and much of it is west of the Mississipi River, yet to be mined. The coal bearing strata on the Cape Breton Island has several workable coal seams between four or five hundred feet apart from each other in the earth. The bottom coal deposit, according to the history of coal explained by geologists, must be millions of years older than the next seam above. Each of these coal seams had to have great epochs of time to form the individual seams. The vegetation which was a potential coal deposit had to be covered with sand and water to form a concrete base between the veins of coal, and then another great forest was accumulated on the surface to form the second coal seam. All of this had to happen in our world to construct all the coal deposits in every part of this earth. Those many millions of years are known as the "Coal Age".

WHAT WE GET FROM A TON OF COAL

1,300-1500 lbs. of COKE

LUMP COKE
- metallurgical - coke
- copper smelting
- iron smelting
- lead smelting
- lead smelting
- iron and steel casting

HEATING PURPOSES
- homes
- industry

CALCIUM CARBIDE
- acetylene
  - chemicals
  - welding

WATER GAS
- heating
- homes
- industry
- chemicals

OTHER INDUSTRIAL
- chemical processing
- lime burning
- beet sugar refining
- Manufacturing of mineral wool

SCREENINGS OR BREEZE
- iron ore agglomeration
- chemical processing
- steam generation

8-10 gals. of COAL TAR

TAR COAL
- carbolic acids
- pharmaceuticals
- cresols
- lysol
- photo developer
- plastics
- phenols
- detergents
- drugs
- dyes
- food preservatives
- perfumes
- rubber chemicals
- weed killer

TAR BASES
- pyridine bases
- antiseptics
- disinfectants
- paint thinner
- pyridine
- clothes waterproofing
- sulfa drugs
- synthetic vitamins

NAPHTHALENE
- insecticides
- fungicides
- plastic dolls
- explosives
- moth balls
- synthetic fibers

... 3 gals of LIGHT OIL

BENZENE
- synthetic fibers
- nylon
- aniline dyes
- food preservatives
- motor fuel
- plastics
- synthetic rubber
- tanning fluids

TOLUENE
- antiseptics
- fingernail polish
- printing ink
- saccharin
- TNT explosive
- aviation gas
- detergents

XYLENE
- motor fuel
- gasoline solvents
- herbicides

SOLVENT NAPHTHA
- rubber solvent
- electrical insulation
- linoleum
- varnish

...5-6 lbs. of AMMONIA

AMMONIUM SULFATE
- Fertilizers
- chemicals

AMMONIA LIQUOR
- fertilizers
- explosives
- household ammonia
- refrigerant
- nitric acid

DIAMMONIUM PHOSPHATE
- fertilizer
- fire retardant

...9500-16500 cu.ft. COAL GAS

HEATING PURPOSES
- homes
- industry

SULFUR
- fungicides
- insecticides
- sulfuric acid

CYANOGEN
- cotton finishing
- dyes

PART TWO

"The Collier's Geology"

Part Two

"The Colliers' Geology"

Coal is a part of the earth's crust that many geologists have named sedimentary rock, and hundreds of these illustrious persons have made the origin of coal a lifetime study. Nonetheless, no factual evidence has proven how coal was developed from vegetation into a mineralized energy potential under the ground. In fact, scientists have mildly opposed each other's theories on where and how vegetable matter was bituminized in the ground. There was at one time a speculation among several of coal's researchers an assumption that the compound of vegetable substances which was later to form coal in the earth's crust drifted as a pulpy mass into regions that are now mining districts. However, that hypothesis was not widely supported during arguments on the origin of coal, and it is generally assumed that the materials which formed coal, decomposed and mineralized where they formed and are now located under the earth's surface. Coal is listed in two general categories, bituminous (soft coal), and anthracite (hard coal), but there are other forms of coal taken from the ground.

The average coal miner's information on geology is limited to a study of mining handbooks, and a primary review in the colliery night schools, and most of the material written into this paper was obtained in a similar manner. The writer's mining experience was acquired in several mines. In later years he had the time to research books at libraries in Canada and the United States. The Random House Encyclopedia has stated coal is the general name for a large variety of fossilized, amorphous, carbonized vegetable deposits with varying percentages of carbon, water, volatile and hydrocarbons laid on the earth's surface during an Age in time called the Carboniferous Period.

The earth we live on is mostly nonmetallic elements which are covered by rocks of different natures. The examination of these rocks has led geologists to divide the immediate crust below the earth's surface into two divisions which are known in mining as stratified and unstratified stone. The stratified rocks have an appearance of having been deposited in layers, one above the other; while the unstratified rocks occur as amorphous masses having no form or direction. It was told by geologists who have made an extensive study of the earth's crust that in mines where the stratified strata has sharp inclines that it was caused by movements of the earth. It is also written in the pages of coal's history, that the sandy sediments which covered

A B B1

STRATIFIED AND UNSTRATIFIED ROCKS AND FAULTS

Beds of sand were deposited over large areas, and there have been many variations of the first surface of the crust which caused the strata to be inclined at different angles, although most of the materials in the crust were originally deposited in a position more or less horizonal. In most of the earlest rocks which were formed in solutions which contained mostly water vegetable and animal fossils are found.

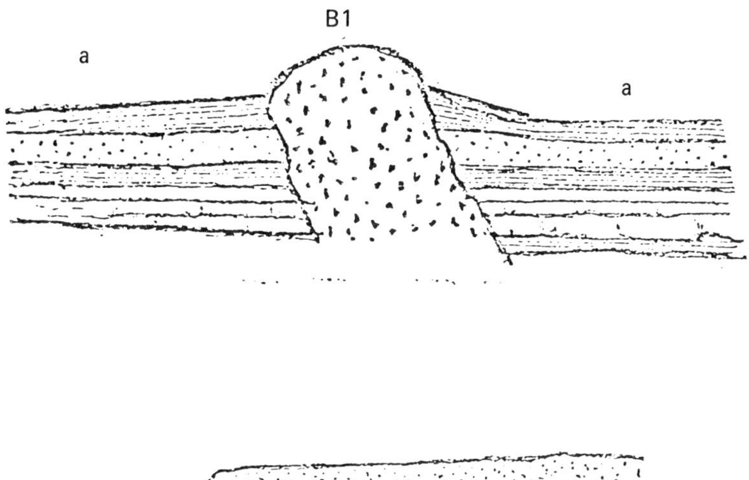

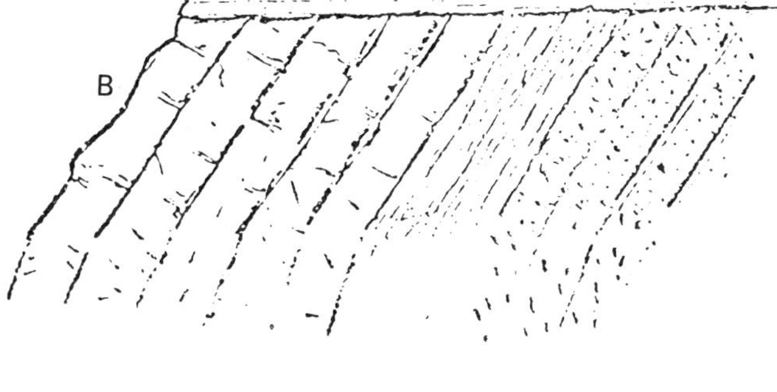

The sketch of the curved strata is known to miners as a "saddle-back fault", in which the stratified rock bends downward and upward in opposite directions, but when the stones dip towards the line of depression the "so-called" "trough fault" is formed in the underground strata. "A" is the "saddle-back", "B" the "trough".

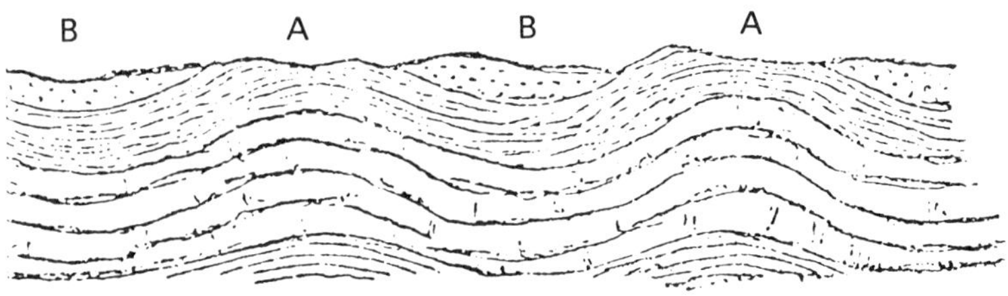

parts of the earth where coal was formed were flat lands during the "Coal Age Periods".

Thirty years ago several of the world's greatest petrographers started petrological investigation of the Sydney coal basin in Nova Scotia. Some of the world's greatest scientists gave systematic classification of the rocks on the Eastern coast of North America. However, the general discourse became the establishment of a uniform classification of the different ingredients found in coal under the eye of a microscope. It was well known by these people that both chemists and geologists have disagreements from sketchy images outlined on former microscope illustrations.

Geochemists know the nature of the rocks below the surface for approximately five miles of depth from drillings and stone samples taken from deep mines, which are generally analyzed in government laboratories. No person has yet, mechanically or otherwise, removed any stone from the earth's crust below five miles, although men have looked over rock and other materials from below that distance mentioned after violent earthquakes.

The coal-bearing strata of Eastern North America is sandstone, shale and limestone, which also has proven oil deposits. Sandstone is well sorted, generally favoring a porous nature, but occasionally filled with silt and clay. Some sandstone over the coal beds is massive and firm with a conglomerate of other materials, such as hard pebbles of shale and limestone. The thickness of sandstone and its strength over a coal face will make a difference in the operation of the "Longwall Method of Mining" in coal pits with sharp inclines. Longwall coal faces advanced several hundred feet beyond empty wastes are in trouble when the permanent strata weight has not broken the sandstone roof above the mined "gobs".

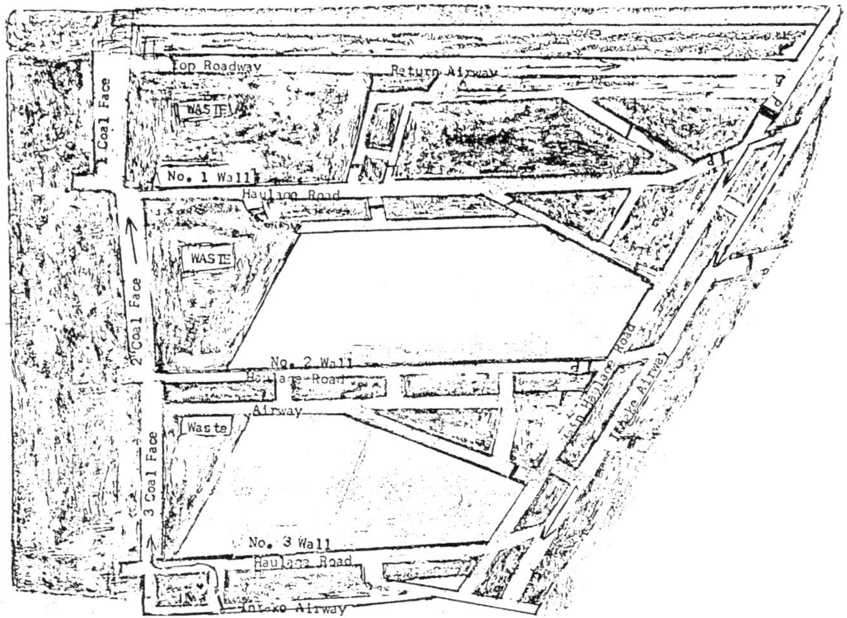

The "Longwall" mining method, sketch of the ventilation system, doors and stopping guiding the air into the coal faces and returning from the workings. The plan also shows the "wastes" and the "haulage roads" in IB colliery at Glace Bay twenty five years ago.

The shale ceilings in the Cape Breton mines are usually a dark blue color of shiny brittle materials which make an ideal operation for the "longwall method" of mining under heavy roof conditions. The weight of the strata over the coal faces is released from a great lot of pressure when empty spaces in the wastes, where the coal was removed, collapses over a large area and fills "the wastes" with stone and rubble from the strata. During "room and pillar" methods of mining, a shale roof may have problems for the miners and the operation; it could be dangerous, and it was up to many miners of the past decades, before the electric lamps improved the collier's vision, to examine the overhead stone strata. Large rocks shaped like bells or old kitchen pots clung to the shale ceiling after the coal was removed and were too often not detected by the workers and later fell without a warning to the floor of the mine. A consolidation of clay, mud and silt make up the structure of shale stone which is occasionally fissle from thousands of centuries under water. Shale is probably the oldest rock which traverses the coal measures.

Limestone is an impure rock which has several elements in its construction, however, during periods known as Pennsylvanian, some pure limestone was found by geologists a century ago. Generally limestone contains many marine fossils, although it may not have been under water as long as shale or sandstone. The creatures which formed the fossils in limestone may have crawled from deep water to feed on shell fish and other animals that lived on the shores of lagoons and died there among the sea shells that are a part of a limestone rock.

Vegetable matter which was carbonized in the earth and later became coal is found in many parts of the earth's crust. During the periods which vegetation was transformed into bituminized mixtures, hydrocarbons replaced the vegetable forms that cannot be seen without a microscope. Longer than the ordinary person would care to look into or reflect upon, nature and time have recurring actions that formed each of the coal seams that are now located in the ground, one on top of the other, but generally a few hundred feet apart. It must have taken many millions of years for the several coal seams to form that are located in the earth's crust in the colliery districts around the world.

Coal consists of plant life which prospered on this planet before people were around to record its history. But the story of coal is mostly a theory, and most people think coal is mainly wood, which isn't correct. However, the exact nature of coal before its formation in the earth may be discovered someday by the combined efforts and the intelligence of chemists, geologists and botanists.

Microscopic examinations have shown that the general substances of coal are leaves, twigs and fruits that dropped from luxurious plants which are now extinct. At one time it was thought that all the coal deposits in the earth were derived from the same materials, or a "mother lode"; this was a misleading theory and most mining students know that each of the coal seams has characteristics not related to the others. At the same time it is also known by an analysis of two pieces of coal taken from the same seam that they may have different percentages of ash, sulfur and carbon. Coal companies that do not have modern cleaning plants for their mineral tonnages bypass in mines thousands of tons of coal to mine workings with high metallurgical value to manufacture coke for steel corporations. Most of this coal left is to be developed at a later date, although much of it will not be mined and becomes an energy loss to the State. It is true that coal will gain sullage and lose heating potentials from limited mining abilities by operators. Coal exposed to the open atmosphere for several months and the lack of an initiative by people, in quality control departments, faulty transportation methods above and below ground, have contributed to the loss of coal for a fuel consumption in the last several decades.

Pennsylania anthracite coal regions have the purest coal in North America; their condemning may be over, unlike bituminous coal which supports combustion at low percentages of oxygen. Spontaneous combustion is practically nil in hard coal mines, as anthracite coal is ninety percent carbon and gains very little foreign materials from the mines to the consumer.

The collier's mining handbooks support the theory that coal was mineralized from "peat to anthracite", and brown coal (lignite), was formed from wood and shrubs during the late centuries of the "Coal Age". Lignite is an intermediate coal of a brownish black color that was formed between peat and bituminous coal, which was not completely mineralized. Italian miners who have experiences in bituminous and lignite mines have told the writer that brown coal was more "gassy" than "black coal".

Some years ago an English chemist wrote the following lines: "I no longer hold the peat to anthracite theory, and other information on the origin of coal is necessary for the intelligent mining student." Studies were made by many persons, but the lack of proper instruments in coal laboratories for scientific experiments on coal's origin has slowed since the first World War, and most of this information remains a theory.

In spite of that rejection on coal's origin, another English chemist and a doctor of science has written that he heated peat and a complex of carbohydrates (cellulose), for eight hours at a temperature of 340 degrees centigrade, under a pressure of one hundred atmospheres. One atmosphere in practice is equal to a pressure of fifteen pounds per square inch of surface. The doctor's experiment produced artificial coal, physically and chemically identical to the coal taken out of the mines, and he later told other researchers, "using the same materials and the average temperature on the earth's surface, it would take eight million years to form the same product that he produced during his experiment." There is no doubt at all that the temperature of the ground and the surrounding atmosphere in the day of the "Coal Age" was a big factor in the formation of coal, but it must have taken more than eight million years to form the coal seams.

Vegetable matter decays under water in the absence of air, and inside the earth for untold centuries decomposed vegetation was building the coal deposits. The putrified plants laid on the crust's surface for unknown periods of time and were covered with sand and water; later disturbances in the earth's mantle pushed them into the ground. The decomposed plants which were transplanted into the earth's crust were changed into water and carbon dioxide, that left a residue rich in carbon. The earth's geologic age is measured in eras and periods; the Paleozoic Era has seven periods and the Coal Age lasted from the Permian to the Devonian period.

Carbon is known to people since history was first recorded,

and is present in all living matter — animal and vegetable. Under the ground carbon is found in three different properties — graphite, amorphous and diamond. Graphite and diamond have crystallized forms, but amorphous has no formation, although it is the chief constituent of coal and the only one of these three elements that is combustible. Carbon has many functions in the manufacturing and mining industries; mechanical mining machines use "diamond bits" in drills and picks, and carbon dies are used to shape cold steel into hard round bars at the auto plants.

The chemistry of the mine gases is a science which has increased the knowledge of mining engineers and coal company chemists through the difficult study of "organic chemistry". Organic chemistry deals with the carbon compounds which include the substances left in the remains of plants and animals. The elements in the gases contain carbon, and the colliers who become mine supervisors should have an elementary seminar on the carbon compounds in the mine gases. In no way will a mine boss be completely competent in the coal pits if he cannot distinguish from each other the several gases found in mines during emergencies. The simple symbols of the underground gases were the first teaching for miners in the colliery mining classes in Nova Scotia for many years.

A molecule is a tiny substance which contains all the properties in a body that is known to chemists as a "mass", and when broken into parts forms atoms which differ from each other in various amounts of hydrogen, carbon and oxygen. The compound substances of two different atoms will unite in the atmosphere to form gases that miners have experiences with in the workings of coal mines. There are only three distinct gases that are the general concern of the underground worker — Methane ($CH4$), Carbon Monoxide (CO), and Carbon Dioxide ($CO2$). Water with a common $H2O$ molecule is a part of these gases, however, it's a gaseous form of $H6O$ in snow, rain and atmosphere smog. The mine gases will be fully looked over in another part of this book, but it should be mentioned here that the mine gases will become poisonous in specified percentages in the atmosphere.

The earth is surrounded by an atmosphere of air which contains approximately twenty gases, and oxygen and nitrogen are its chief elements for about six miles above the earth's surface, although there is some thin air a hundred miles up in the heavens. The reader may notice the absence of carbon in the atmospheric gases, but traces of carbon dioxide in the surface atmosphere is important to the growth of vegetation on top of the earth's crust. "Pure air" may be found at the center of the world's two largest oceans, and it would contain twenty percent oxygen and seventy-nine percent nitrogen; both of these gases will slightly exceed these percentages, and the remaining elements will contain important fragments of carbon dioxide gas. Nonetheless, it is possible that pure air will not be found in any part of our world.

The atmosphere above the earth's crust has air currents which pick up impurities as they travel around the world; this changes the composition of air in most all environments, especially on the lands that people inhabit. The earth embraces pressurized elements which are forced to the surface in the form of gases, which will lessen the content of fresh air as it passes over the earth's continents, but industrial towns and cities are the worst offenders. The rotating mass of materials which forms our earth will drag winds with its movements into warm climates around the equator which expands air into a vapor above the earth's crust, although the earth's inner materials pull things towards itself with gravitation. However, the rotating mass also creates an outward push which is known as centrifugal force, which is greater on the surface, at the center of the earth. The centrifugal effect at the equator will push foul air into latitudes in the atmosphere, then the atmospheric pressure and winds will produce new air in many environments.

Fresh clean air is necessary in mines for the good health of the people who are employed underground, and the composition of air also changes as it circulates through the workings of coal pits. Expired air moving out of mine airways and out of mine return air shafts into the surface atmosphere will have a loss of oxygen, nitrogen percentage changed and the carbon dioxide increased. In large well-managed mines "air samples" are taken several times a week, but there are hundreds of mines in North America that never have an air sample taken of the dust and gas that miners are breathing at work underground.

Expired air with 16.03 points of oxygen, 79.47 points of nitrogen, and carrying 4.50 points of carbon dioxide will not support combustion, yet it may not be poisonous to breathe. Nonetheless, people should not remain in the environment. That atmosphere can change without a warning and become poisonous when the oxygen content drops to seven percent, and the nitrogen moves upward to 93 points. It would be impossible to light the flame safety lamp in the environment just mentioned here. Neither will the safety lamp burn in 17.2 parts of oxygen, 64.8 parts of nitrogen, and 18 points of carbon dioxide in the atmosphere. The high percentage of carbon dioxide in the air "blacks out the flame". However, the lamp may burn with the same percentages of oxygen and nitrogen which are mentioned in the last sentence, but the carbon dioxide would be changed to a small percentage of carbon monoxide. Carbon monoxide is seldom found in mines, but it is extremely dangerous if inhaled. Hydrogen sulphide is another treacherous gas which may be found in coal pits after large fires. Most of the information on the underground noxious gases will be explained in the third part of this book.

The velocity of air underground is measured with an anemometer, which gives speed, force and quantity of air moving in and out of the main airways leading into the mine workings. The quantity of air leaving the mine "intake air shaft" may be over 200,000 cubic feet a minute; that air is divided underground and each section of the workings is alloted certain shares, generaly more than is actually needed for a safe working condition. The first anemometer was made by an Englishman named Wolf. Wolf's anemometer had four sails like those on a "windmill", and was used generally to measure surface winds. It is usually stated in mining books that "Casartelli" improved the mine anemometers, and that he placed a watch in one of "Biram's" anemometers during the last century. Although the anemometer has been around mining districts for many years, it was not always available at the collieries. This tale is an authentic account taken from a mine manager's textbook. From a measured distance in a mine airway, the mine boss and the colliery mining engineer stood at each end of the track, one with a handful of gunpowder that he lighted, while the other person timed the smoke as it passed the measured distance.

The "old underground furnaces" were instrumental in sending thousands of cubic feet of air into the mines, but many of the first mine bosses did not know or have the instruments to divide it correctly in gassy workings. Although lighting any material underground is and was a dangerous practice, even during the days that miners were supplied illumination with open burning fire torches, these old mining people thought it was a safe practice for experienced mine officials. "And they would never allow the 'run-o-mine' worker to perform the task of measuring air by exploding any kind of devices." Mining history has revealed that the lack of oxygen in the air at the coal faces during the first days of mining was responsible for many of the mine explosions not taking place at the "room-ends" (coal faces).

Methane has a close association with vegetable matter which became coal, and only a very small percentage of the other gases mentioned throughout this book is released from the coal deposits. Methane is also known as "marsh gas", and it may

This sketch shows a large fissure in the stratified strata inside the crust of the earth which holds gas, oil and water. "C," on the bottom is water. The center of the fissure is oil – "B." "A" at the top is gas.

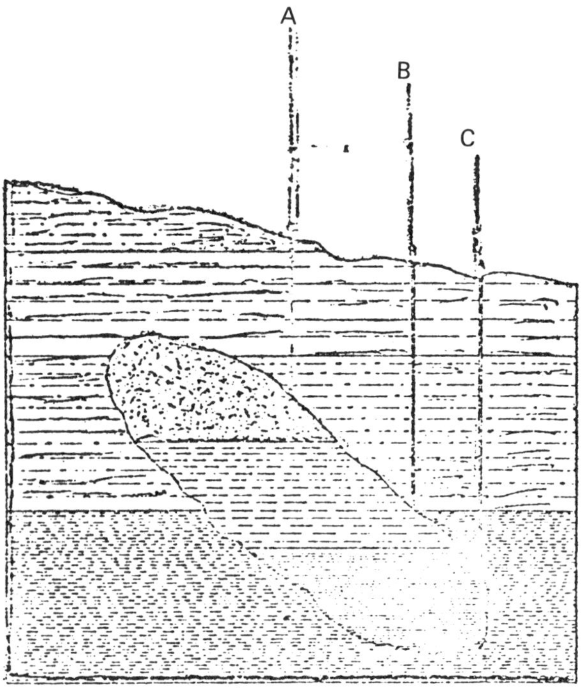

be found in lowlands and boggy marsh fields just before "sunrise". During the Carboniferous Period, the earth was a soft warm soil of organic construction, which scientists claim formed coal and methane gas under the ground. Although some others have written, the atmosphere above the earth's surface contained great amounts of methane which is now found in the rock strata. During prehistoric epochs of geologic time the earth's vegetation developed into vast germination, which later fell upon the soft surface of the crust. The atmosphere surrounding the earth at that time was a thick vapor which turned into water, and covered many parts of the earth's surface for millions of years. The water washed sandy sediments over the decayed vegetation on the ground. The marsh lands, which were in the vicinity of the great forests and flowers which formed the coal seams, were also buried under sand and water. Disturbances in the earth's mantel locked the marsh bogs in the crust, and in another era which took millions of years, methane gas was formed in the ground, and often adjacent to the coal deposits.

In most bituminous mines small reservoirs of methane are located in the stone above and below the coal, and this gas will drain through fissures in the strata into the mines' atmosphere. It is the opinion of the majority of mining people that the methane gas in the stone which surrounds the coal beds was pressed out of the coal by the tremendous pressure inside the earth. However, the coal seams contain pockets of methane which caused small expulsions of pure methane. Although I have never seen that happen, I did have an experience with a large ejection of methane gas from the stone strata at IB mine in Glace Bay. That gas flowed at explosive percentages for two hours into the mine return airways, and the pressure from the gas was pushing back approximately 50,000 cubic feet of air.

The most noted account of gas expulsions in coal mines that this writer found in old mining manuals, resulted in a tragedy at a small colliery town outside of Mons, Belgium. Violent gushers of methane gas filled the underground workings of the village coal mine at a great pressure. The gas filtered at explosive percentages into the pit's only air shaft, and was ignited in a boiler house on the surface. The entire working force of 121 colliers died in the explosion. It took another mine disaster to stop one-shaft mining operations on the European Continent during the last century, when over 200 persons were suffocated in a one-shaft deep mine after an accident closed the only air shaft for the miners' ventilation. During 1968, in the State of Louisiana, gas was fired in a one-shaft salt mine, and workers died in the blast. The wretched reminder of this tragedy was told by a federal mine inspector who examined the mine a few days before the blast, "I told the mine officials to repair conditions in the shaft."

Both management and labor in the coal industry pioneered safety at work before all others so that coal mines can be worked safely by skilled colliers. The setting for any safety program will be found wanting if there is no coordination from the people who are chiefly related to the subject. One-sided directions and regulations are not the answer to achieve workers' safety, although people who abuse rules and privileges should be disciplined. Miners act no differently than others when penalized for infractions on the job. However, a stern lecture on instructions was generally the section boss's procedure. Nonetheless, when the safety of a mine and its workers is jeopardized by careless bosses, and others, it is time for action to be taken.

Learned people with facts and history of the mine gases have lectured to colliery officials on the dangers of gas and coal dust which form in the underground atmosphere, and is an explosive environment unknown to the workers. The average miner has a fair amount of knowledge of how and why gas is lingering in his working area, but very few of the workers and the mine bosses realize the potential explosive nature of an atmosphere which contains small amounts of unexplosive methane gas and sufficient coal dust which create that condition. It is natural for gas to release itself from the coal face after the workings are mined, and methane gas can be heard issuing from a "cleaned-off" coal face after the miners have left for that day. Methane gas is not poisonous in itself, but in a closed environment it will reduce the oxygen in the air below the necessary percentage to support life or fire. The natural gas taken from the earth is methane, and it will burn and explode in homes as well as in mines.

The competent mine supervisor is much more than a mathematical time-keeper underground; although the miners' safety is his first consideration, he must also produce coal at a favorable cost for mining investors. However, many people think coal mines are above the average industry in compensation injuries; there are mines with safety records that surface employers have not achieved in a lifetime. It is people who make any industry a dangerous place to work. Any coal mine worked without sufficient ventilation is a hazardous place to work. Neither can a mine be considered safe when the workers and mine bosses are pressured into high productions for a few extra dollars, and serious accidents are needlessly caused.

No 3527

Province of Nova Scotia

Department of Mines

This is to Certify that RONALD H. McINTYRE

of GLACE BAY County of CAPE BRETON

now the holder of an UNDERGROUND MANAGER'S

Certificate having duly complied with the provisions of the Coal Mines Regulation Act, is hereby granted a Certificate of Competency as a

MINE MANAGER

Given under my hand and seal of office, at Halifax, this 15th day of JUNE A.D. 1942

R. D. Currie

Minister of Mines.

Mines can be operated safely, but miners should not be working at the faces when known pockets of explosive gas are not cleared from the air.

Geology is a difficult study for most people, and this paper cannot enlarge on the views that the miner has learned in the colliery mining classes and a home study of miners' textbooks. The mine student will attend the colliery mining classes for four Winters, and it must be admitted very little geology was taught in the classes, but the miners talked to each other on this subject and a few acquired books that they studied at home. Now miners can visit libraries, study books on many subjects — including geology — and in the Cape Breton coal fields the miners are attending classes at a college near the City of Sydney. It is not now the work or job of the mine manager or the colliery mining engineer to act as a geologist, but they should have some knowledge on the substances which cause coal to form in the ground. The collier who is attending the practical mining classes should be taught the rudiments of chemistry and geology before he is passed out of the night school classes.

The practical mine student who has successfully obtained a "Mine Manager's Certificate", after at least four years of study while working underground, and writing yearly tests on all phases of work in the coal pits, "must become the master of his destiny", have some labor relations lectures behind him, and an overall understanding of mining methods and also the gases found in the mines, before he is appointed to the position of a mine manager. Nonetheless, that same person had to prove his "metal" on other official jobs underground before his appointment as a colliery manager. Underage boys are no longer employed at the collieries, and most of the young men starting in the mines have high school grades, and enroll in the mining school classes for their advancement in the industry and for the safety of their fellow workers.

Engineers, metallurgists, geologists and chemists are the very essence of coal mining, and this should not be forgotten by production giants in the colliery districts. There is an inclination among mining people to regard these mining "specialists" as paradoxes who use contradictory mining statements. Nonetheless, over 300 years of mining experience has proven that the only efficient colliery management was gained from people whose roots are in the mining districts. It would be impossible to exaggerate the importance of the combination of a college mining engineer, and the practical mining student from the colliery night schools in the mines. Coal companies are paying agents to locate miners capable of holding down official jobs in

"A" The North Atlantic Ocean basin which contains coal, oil and gas.

"B" A limestone overthrust fault shown from the surface, and located in the State of Oklahoma, U.S.A.

A

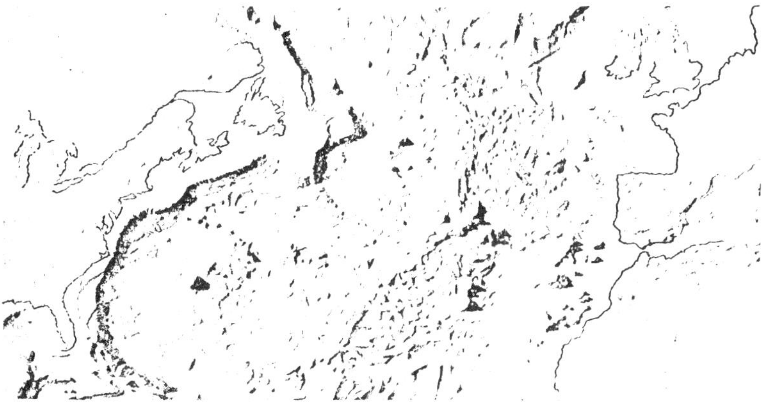

B

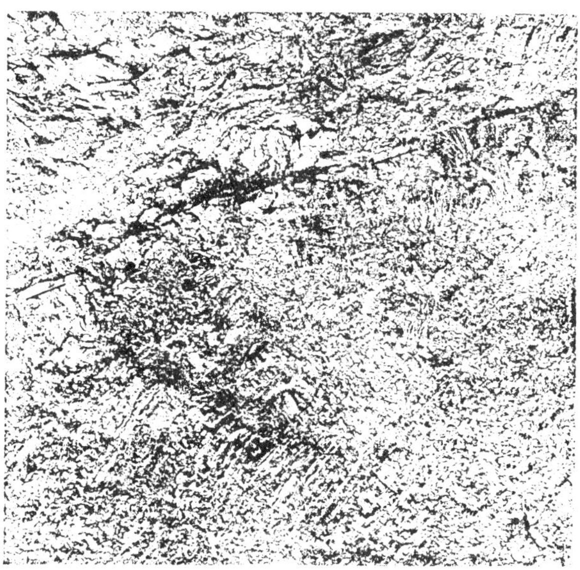

the pits, but people with practical mine school training are hard to come by in Canada and the United States. Even in Nova Scotia, where mining schools were started by a group of colliers many years ago, the mining classes are far below the enrollment when this writer lived in Glace Bay. Yet, there are several hundred potential miners waiting for work in that town today.

Coal companies all over North America know that they will need mining people who were trained in practical mine schools in the colliery districts, and they should put forth a greater effort to have mining classes in their own coal towns. Yes, there are people writing "government question and answer" papers who become mine bosses, but how many of these people would pass an English miner's colliery manager's mining exam? Coal's great future may depend on the collier's mining experiences in the pits, and "book-learning" at home, and the teachers in mining classes in coal towns. "The colliery school student is money in the bank for new coal companies."

Most coal seams slope into the ground at different angles from an outcrop on the surface; this is known in mining as "the dip". Before the coal deposits were set in permanent positions in the earth's crust, other materials foreign to coal's composition were pressed into the coal measures, which formed faults in many coal seams. Generally, mine faults are not connected with known world faults. However, this writer talked with two geologists, who had traced an underground fracture in the earth from Europe, across the Atlantic Ocean, into a mine in Nova Scotia. Coal is a composite of organic vegetable matter, unlike stone which is a direct result of mechanical deposition of mineral matter. The earth was at one time a mass of molten materials; it cooled during the centuries before the crust was formed around the mantle. Nonetheless, the crust at one time is described by geochemists as a "cooked mush" from the smoldering fires beneath its surface.

The stratified layers of stone in the crust of the earth break off at intervals and uproot the coal seams. These mine faults vary in sizes and conditions, and disconnected pieces of stone and coal are mixed in the strata, sometimes level, or vertical and over-lapping into folded positions where the mine roof and floor are reversed. The over-lapped faults may contain gas in large quantities that miners call "gas-blowers", and also true fossils of creatures and plants which were a part of the earth's surface at one time.

Many of the rocks with peculiar shapes and sizes that this writer looked over were not true fossils, just rocks that were shaped into forms by water and ageless periods of time. Some years ago a miner gave me a round stone 10 inches in length and

"UP–THROW" FAULT

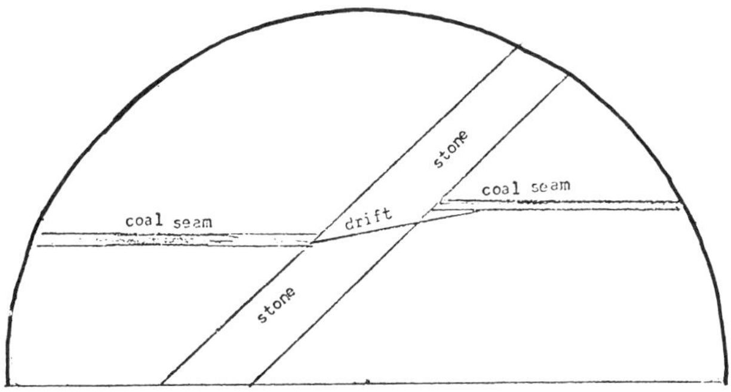

This is an "up - throw" fault, which was found in an old British coal mine. The coal seam was "bumped" upward one hundred and fifty feet from the regular angle through which the seam was found and mined.

one inch in diameter which fell out of the mine roof. The peculiar factor of that stone was that it was perfectly threaded and headed into the similarity of a steel bolt.

Coal, oil, and natural gas are fuels which formed in the earth's crust from plants, bogs and animal remains, and all three are considered fossil fuels which contain carbon and hydrogen gases. Most of the raw coal taken from the ground has various percentages of sulfur, which retards the cementing of coal into "coke" used in the manufacture of steel. Coke is a dark gray, brittle, combustible material of a porous nature that is transformed from bituminous coal into approximately 90 percent carbon, and used as a fuel and a reducing agent in the chemical and metallurgical processing of steel. Coke is made in a closed heated chamber which draws off several of the gases in the coal and leaves mostly carbon. The heating vessel is known as a "coke oven". Resin is an organic substance in coal which forms the cementing materials in coke, and the best metallurgical coal seams will have a fair amount of resin bodies.

Coal will become a leading source of energy before the end of this century, in North America, but skilled miners have to be in the pits to save the industry from great losses of coal that most people never hear about. Thousands of tons of coal were left in the underground wastes during the first years of mechanical mining. Today, thousands of underground acres of good coal that could be cleaned in modern wash plants for thermal power is bypassed and may never be mined and will become an unknown loss to the State. Coal was deliberately pushed into the "gobs" to start another shift. "Gob coal" is seldom removed from the wastes; when left there, it will later heat to a high degree of hotness and form a smoldering fire in the wastes. Some of the gas found on levels above coal face walls leaked out of fiery wastes.

The mine bosses have very few ways to stop a smoldering fire in a closed longwall waste. At first, water should be run through the waste stones; if that does not put out the fire, they would have to dig through the waste to put it out. The large mines presently have underground fire walls, and it is possible to flood each section separately with water to black-out a fire. However, most coal pits would have to be wholly flooded in cases of stubborn fires. The safest method to stop mine waste fires is to have underground supervisors who do not allow large amounts of combustible materials to be placed into longwall gobs or wastes.

Coal mining is a difficult study and hard work that people from the mining districts have inherited. The colliery mining classes and textbook study were started in England during the 1800s, but the British colliers have been continuously mining coal for nearly 1,000 years. Eight hundred years ago, Henry III of England prosecuted his subjects for burning coal, although Britain later became a great power through coal, on land and sea. The British coal trade was world-wide, and her ships carried coal to every port in every land. Three hundred years have passed since coal started to change the habits of people around the world, but it was not until late in the 1800s that the black mineral became known as "King Coal".

The underground mining boss who has several years of mine experience and two or three years of practical mining study should be proud of his safety record and explain to others his principles of mine safety. The colliery mining schools in England and Nova Scotia have taught many young men to become engineers, and miners with mine manager's certificates.

The American mine inspectors are under-manned and often over-worked, but they have not spared themselves for their colliers' safety in the mines. These people need help, and it has to come through the establishment of practical mining schools in their coal districts. The Mine Inspectors' Institute of America has, at its own expense, distributed mine safety literature from

No. 2449

Province of Nova Scotia

Department of Mines

This is to Certify *that* RONALD H. MacINTYRE

of GLACE BAY *County of* CAPE BRETON

now the holder of a FIRST CLASS MINER'S

Certificate having duly complied with the provisions of the Coal Mines Regulation Act, is hereby granted a Certificate of Competency as a

MINE EXAMINER

Given under my hand and seal of office, at Halifax, this 22nd *day of* JUNE

*A. D. 19*38

Michael Dwyer

Minister of Public Works and Mines.

mobile units to miners in the American mid-west, some years ago. Why not use mobile coaches and teach miners two or three nights a month in the coal towns? The classes could be taught by retired engineers, inspectors and experienced coal miners. The mobile units would have printed mining discourses for the colliers to study at home, but the classes must be schools which teach, not just bookmobiles. However, the project would have to be supported and supervised by the Government.

The coal industry lost many of its best customers to the gas and oil people a few years before the second World War. The oil corporations promoted a liquid product that they claimed would not pollute the atmosphere, although most people today know the combustion from any burned material will defile clean air. The large ocean ships were converted into "oil burners", and coal companies were experiencing their first great loss of revenue. Then railroads reneged on contracts with coal corporations, yet coal companies kept them in business since the days of "pokey locomotives". However, steam locomotive firemen were complaining for years about the "poor grade of coal" they were shovelling into smoldering fires. Nonetheless, coal companies could not mine any grade of coal to compete with the low grade of oil that railroads were using to operate their new locomotives twenty some years ago. However, the miners, and the quality control people at the collieries, and other mine officials must share the blame for the loss of revenue when city people stopped buying coal or its products. During the 1950s a great majority of the people in North America were thinking coal was an ancient energy and soon would be forgotten. The United States has produced 700,000,000 tons of coal during 1979.

"Run-O-Mine Coal" was generally sold to the people in the colliery districts; many complained —mostly persons not in the company employ — but the company blamed the miners for the large amount of stone in each ton of coal. "Run-o-mine" coal was run through "picking belts" and people removed foreign materials from the coal as it passed by them on the "bank-heads". The coal exported from the colliery was passed over large screens, and often graded, but there were a few coal companies after the "fast buck" and shipped run-o-mine coal to their customers. Forty years ago, large coal consumers relied on past performances for their next shipment, but many were disappointed and oil became popular with coal customers. The coal seams will contain approximately only five percent of impurities. That percentage would increase before the consumer handles it from modes of mining and inadequate quality control organizations at the mines. Most coal corporations presently have wash plants near the mine shafts, but some prosperous companies have plants where the coal is "washed and dried", after which it is pulverized into products for commercial use outside the heavy industries.

Coal's great future has to be in thermal plants, and the manufacture of coke for the steel industry. Electric power plants fired by imported oil in Canada and the United States have a tremendous cost which coal will reduce. Environmentalists have lobbyists, at the door of government offices, who wish to drain the last barrel of oil from the earth, and thousands of people in colliery districts have no jobs.

Corporations must have profits to operate successful industries, and sensible mining people know from experiences that high costs are created in mines by outdated mining methods and equipment. Unlearned fire-bosses confuse workers and a few have several minor accidents before they have a major misfortune underground. Some coal pits make profits for months, and then have to shut down for weeks through carelessness which caused fires, blasts, and too often, deaths. The final count for unskilled mining bosses who allow gas to linger near the coal faces and other nonchalant regards to mine laws means unemployment in the coal towns, and old miners moving slowly into their homes on canes and crutches. There are miners living today who worked 30 years in coal mines who never had a pension from coal companies, because they worked for companies who would not pay them an old age benefit, nor did they pay fair wages when they were paying big dividends to stockbrokers. The coal demands are growing greater, however, coal corporations have to watch the steel markets, or they will be in over their heads and may be drowned in over productions and unemployed workers.

Coal may never again be "poked and cursed" in kitchen stoves, but the products which coal produces will supply the power needed in homes for many future years. Coal sales people may be women who will display coal energy and its products from a chemist's analysis, and the few women now working in the American coal pits should be looking for these jobs. Mining is a challenge for males with more muscular development than famales, although most of the work underground is not "pick and shovel". One of these women told us she was using a shovel, and was expected to do the same amount of work as the men.

Neither coal nor oil are replaceable in the ground. Corporations and governments are drilling "dry holes" and looking at others which produced millions of barrels of crude oil, much of which was wasted and lost on the ocean sea lanes. It is now believed by many people that oil companies are concealing the amount of known oil deposits in the earth's crust. Nonetheless, oil will be around for several more decades, although some of it should be left in the ground for future emergencies. Natural gas will relieve the consumation of oil for homes, but it will go out of existence for commercial use during the next century. Although methane gas is not poisonous, it would be harmful in a closed kitchen as it reduces the percentage of oxygen necessary to sustain life in unventilated rooms. Natural gas will burn and explode from tiny sparks of fire when gas jets are allowed to leak into closed rooms in the home. Commercial gas is manufactured by distilling bituminous coal; it is methane. It's not as pure as natural gas, but just as explosive.

Coal tar is a sticky black liquid obtained when coal undergoes distillation to manufacture gas or oil. Medicines, dyes, nylon, plastics and many other useful products are made from coal tar.

Oil and gas will not be problems for the people seeking energy in the future centuries; there will be none left at the rate of our consumption, and the lack of conversion methods, before the end of this century. However, long before nuclear power is proven or solar heat and the substances from the hot mantle are harnessed to supply energy for factories and homes, coal miners will be in the pits trying to supply the energy level we now enjoy. Atomic power will never become absolutely safe, but it will be used in the future.

The Nova Scotia government has supported the colliers' mining school for over fifty years; they paid teachers and supplied mining literature for the students. During World War II, the Canadian federal government established a mine training program. Young men from the coal mining districts were "hired" at the mines and trained by older miners of retirement ages; later the young miners were allowed to work in pairs at the coal faces, however, not before they passed oral examinations for miners' certificates. It is a mining law in most coal mining countries around the world that miners who engage in the actual mining at the coal faces have first or second class "miners' papers". The examination for miners' certificates is oral, and is generally taken before a Board of several authorized miners and the mine inspector for the district. However, such Boards should not have the authority to issue "fire-bosses" certificates to miners. I cannot doubt, in view of evidence given before mining commissions, and other experiences, that

1977 WAGE SCALE FOR WESTERN CANADA'S MINERS

WE NEED PEOPLE LIKE YOU

If you're a mature person, fit and like to do a good day's work, we would like to hear from you. Starting wages at the Flin Flon or Snow Lake operations are:

Journeymen Miners	**$7.28 per hour**
Experienced Miners	**$6.55 per hour**
Mine Labourers	**$5.915 per hour**
Plant Labourers	**$5.75 per hour**

Miner Apprenticeships are available to employees in our underground operation.

Previous experience in Mining, Smelting or Refining an asset but not essential.

The company's benefits program is one of the best in the Industry with Dental, Medical and Drug Programs for you and your family—without cost, and attractive vacation and vacation allowance plans.

Our operations are located in the great fishing and recreation area of Northern Manitoba. City facilities include a modern Hospital, Schools up to and including grade 12, Shopping Areas and a wide range of Community Activities.

Our representative, Mr. Gerald Nomeland will be conducting interviews at the Canada Manpower Centre in Bridgewater August 17-18. Please contact the Canada Manpower Centre nearest you.

unschooled mine management has allowed mining laws to deteriorate underground, even though mining people and the mines are equipped with modern safety devices, and the most modern mining machinery which the world has ever known. The first question asked nowadays after a serious underground explosion: "was the mechanical methane detector working?" Maybe it was, but who was there to inform others in the section that the atmosphere was explosive? Has the cost of coal become so high that underground supervisors with flame safety lamps are not around gassy workings anymore?

The future colliers do not wish to expose themselves to the hazards created in mines by unlearned mining bosses, although economics places people on jobs they do not like or know the problems on which they have been started. That statement does not mean that people cannot become skilled and content after some training on the job. Thousands of people never had the kind of positions they dreamed of or wanted, but they were proficient and happy at their work.

However, we live in a free country, and the average citizen with a high school education and some thoughts of security is leaving the coal fields. Most of the boys and girls leaving small industrial towns and villages are moving into the large cities to become other persons on unemployment rolls. These young high school men are needed at the collieries back home, and many of them will remain there, in spite of the high unemployment in coal towns, when they know the local colliery is operated in a safe manner by their own people who were schooled in practical mining classes in the coal districts.

PERFORMING during the 1978 St. F.X. Homecoming activities scheduled for this weekend will be The Men of the Deeps, North America's only miners' choir.

PART THREE

"The Miners Flame Safety Lamps"

"THE FLAME SAFETY LAMP"

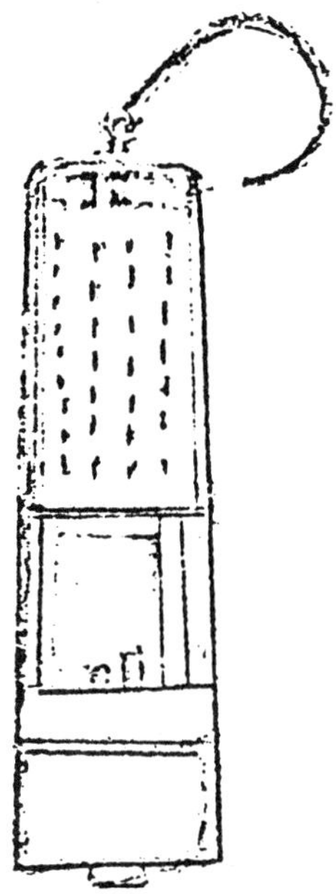

Part Three

"The Miners' Flame Safety Lamps"

Foreword

Numerous theories were expounded through the eighteenth century in attempts to place safe illumination in the gassy underground atmospheres where miners extracted coal from the stone strata. It was not until some information of the substances contained in the mine gases was discovered that safety lamps were invented for the miners. Three people showed mine lamps in England during 1815, all at different times, but in the same year. The three lamps designed at that time, 1815, were distinct from each other. However, two of the lights used the same principle for lighting, although years later it became known that there was a slight difference in the functions of both lamps.

Historians have written, "Davy invented the safety lamp for the coal industry in the year of 1815," nonetheless, another capable person claimed the invention and the priority for the mine lamps. Even though only a few mining people supported George Stephenson's claim at that time, this article will tell that he was the first person to build and use a practical mine safety lamp. The third lamp which was shown in England was made by a colliery medical doctor, and it became a popular light with the rank and file at the collieries for many years.

The general purpose for writing this article is to describe the functions of the miners' flame safety lamps. The history connected with lighting in coal mines before and after the safety lamps were invented is an interesting story. The writer's experiences in mines, and the several Winters that he attended the colliery mining classes in Glace Bay, Nova Scotia, and the mining legendary found in books at libraries throughout the United States and Canada, will be the feature of this section.

Mining reference books were at one time published in the United States, and sold in coal towns all over North America, but they are no longer found in the sidewalk bookstores. Strange as it is, British mining literature for people from the coal fields may be found on the shelves of libraries in the large American cities far from the mine shafts. The English collieries mining books were always considered the most informed discourse for the miners who studied and advanced themselves in the coal industry, and now they can be purchased through the mail from the United Kingdom. At the same time, it is well for students using old mining manuals to check the writing as some notes were improved upon in later years.

England was the first country to form mining methods in the earth's crust, and much of the discourse written on coal mining around this world is a copy of the British mining literature for the workers' safety in the pits. However, during most of the 1800s discipline was strictly enforced to increase coal production from the mines, and thousands of colliers died in the pits because safety was not meted out in the same manner. The mine owners or agents threatened the mine bosses, and they turned their resentment on the backs of the underground workers. Serious accidents had little or no cost for these mine owners, as no compensation was paid for injuries or fatalities at that time.

During the 1700s the British government appointed commissioners to report on the development of the coal seams. They also reported on housing and conditions in the mining districts. It is possible that one or two of these commissioners or agents were in Nova Scotia during the early 1800s, and left some written records of the colliers' state of being at Sydney Mines, on Cape Breton Island, over 150 years ago. The first miners and their families in England were housed in coal company-owned buildings, and did not have to pay rent. In Nova Scotia the early miners were given free coal to heat their company houses, but a small amount of money was deducted each payday from their wages for house rent. Two or three of the first company houses are on exhibit at the Glace Bay Miners' Museum compound.

Although the coal mine regulation act was enacted in England in the middle of the last century, there was no significant reduction in mine fatalities until the colliers' mining schools were supported by the government. Most of the British coal mining personnel who learned mining methods during this century were started in their colliery mining classes. It is a fact that wherever coal is mined in any part of this earth, "a miner with an overman's certificate is sure of a job."

The weight of the world's atmosphere pressing on the earth's surface forced air into the first mine shafts which ventilated shallow mine workings in a fashion which slowed the advance of mining methods underground. Methane gas issues from the mine strata after the coal is removed from its original location, and its displacement could create a flammable atmosphere in the immediate surroundings when new air is not moving into the coal faces. The lack of a steady air pressure in "room and stall" methods of mining coal lessened the size of the pillars in the first mines to hold the weight of the overhead stone strata.

The very first mines were holes in the ground a few feet

under the surface of the earth, and when the gas became a nuisance to the workers, another pit was dug out of the hard rock in the ground. Later, the underground laborers were forced to mine further inside the earth's crust, and then the gas became a deadly problem for the people who mined the coal faces. Fire torches, that each miner carried underground for illumination, could not be used at the coal faces when the "room-ends" were gassy, and quite often the miners at the coal faces were working in the darkness. Generally, the burning torches were left approximately 50 feet from where the miners were digging and loading coal, when flammable percentages of firedamp collected at the "faces" after blasting in the rooms. It is noted in mining history that some miners hung dried fish skins on the "room-ends" for reflectors; the fish scales gave off a faint phosphorescence that the miners used to survey directions in the darkness.

Small tallow candles and open oil lamps replaced the torches for the miners' illumination in the 1700s, but open oil lights were used in the North American mines all through the 1800s and periods during this century. Then, too, we find coal companies which used closed lamps and open lights underground in gassy coal seams. The miners at the coal faces were given the flame safety lamps, and the people who transported the coal to the "pit heads" used open flame lights. The most noted mine lighting apparatus before the safety lamps were introduced in the English coal mines was the "steel mill", and it was in gassy mines for more than half a century. The history of the steel mills will be told in other pages of this book.

At one time mining people separated methane and firedamp as different gases. However, this book will not distinguish them from each other, and will treat them as one and the same gas. Methane is a light gas which intermingles with air and other gases. In spite of the known fact that firedamp may rest on top of a layer of heavy carbon dioxide gas, it will diffuse in time and form another gas.

Firedamp is a mixture of gases, mostly methane, accompanied by small amounts of nitrogen and carbon dioxide. Methane gas mixed with different percentages of oxygen in the mine atmosphere is generally known as a body of gas or firedamp to the mine officials. The amount of oxygen and methane in a body of gas in the mine workings will determine the explosiveness of the surrounding air, which increases and decreases in different mixtures of the atmosphere. Methane has several important explosive stages after it forms a body of gas, as it is a known explosive substance. However, it is also non-explosive in high percentages in the mine atmosphere and will not support combustion. Later that same body of firedamp will become flammable when the oxygen increases and the methane decreases in the atmosphere. A collected mass of methane in an underground workings of coal mines must have approximately 95 points of air before it becomes flammable, after which it burns fiercely in a fired mine atmosphere. The flammable gases found in coal mines are methane, carbon monoxide and hydrogen sulphide, and all three have what is known as ignition temperatures, but each must have different percentages of air to burn or explode. Methane and carbon monoxide have the same ignition temperature, 1202 degrees Fahrenheit, while hydrogen sulphide will explode at the low heat of 750 degrees Fahrenheit. However, it is proper to state here that carbon monoxide and hydrogen sulphide are seldom found in mines, but are watched for after mine explosions and fires.

Methane gas has a high and low explosive limit, and any combinations of firedamp between these confinements in the regular mine air is explosive gas, but with different explosive forces. The explosive ranges of methane gas are between six and 16 percent of firedamp. The gas becomes fiery in the mine atmosphere and explodes inside the flame safety lamp. At 10 percent in the regular underground air, methane has its greatest explosive force; when the same body of gas has a reduced percentage of six parts of firedamp it burns inside the safety lamp with security to the mine examiner. Methane in a near pure state will black-out the flame in a safety lamp.

Although the gases found in mines are comparatively few in number, they are generally poisonous and flammable, and the average miner should have some knowledge of how they occur in the pits. The mine section boss must have a thorough insight on the properties and behaviors of the gases before they become deadly atmospheres for the people that he is responsible for in the workings of coal mines.

Firedamp is searched for in the mine workings with gas detectors and flame safety lamps, but the real watchdogs at the collieries are the mine deputies (trusted people), who use the flame safety lamps at their work in a night examination of the innermost recesses inside coal mines. The mine safety lamp burns and explodes methane gas with security inside the locked vessel during an environment examination which contains explosive firedamp. Generally, the fuel flame is blacked-out in a gassy atmosphere, but the gas will burn inside the steel gauzes until the examiner removes himself and his lamp from the immediate environment. When the safety lamp is placed in a mixture of six percent or less of firedamp, the gas burns a halo (a

"ILLUSTRATION"

Methane gas with a percentage of six percent or less in the mine atmosphere will burn in the form of a halo inside the miners' flame safety lamps, and give the percentage of fire-damp in the immediate mine workings before it becomes explosive.

Caps over Oil-Flame. C:

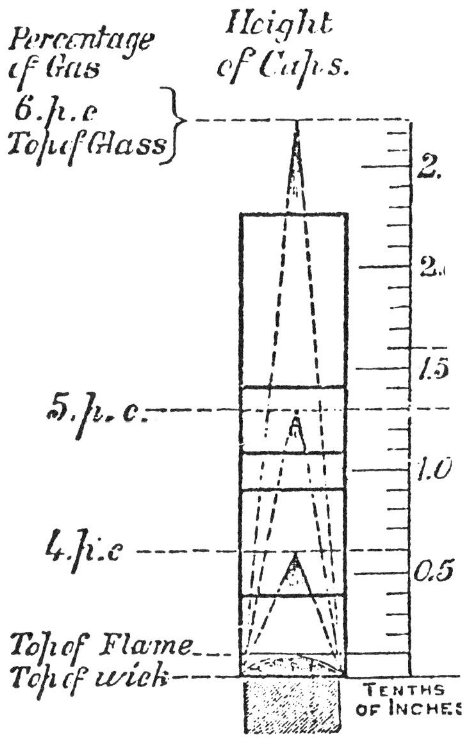

The tip of the cap is 0·2 inch above the

Fig. 581.—FLAME-CAPS SEEN

gas cap) above the lowered fuel flame inside the light's casing. The mine deputy "eyeballs" the height of the gas cap, and each inch above the tiny fuel flame tells him the approximate percentage of methane in the air.

Gas detectors have been around mines longer than most people realize, although it took a lot of years before the instruments became common in the pockets of mine officials. It is possible that the gas detectors carried by mine bosses will not give the correct reading of the percentage of firedamp in the underground atmosphere, as this gas contains modicum amounts of other gases. One of the first mine gas detectors was made by a music professor in Europe; it was supposed to let out musical notes when placed in gassy air. This writer has taken mine gas samples into laboratories that showed the same percentage of methane that the mine deputy had written into the mine log book from his reading with the flame safety lamp.

The gas cap is visible inside the casing through the thick safety glass in the modern-day mine lamp, until the mixture of methane gas on the exterior of the light's casing has diffused in the atmosphere.

The flame safety lamps used presently in the coal pits have steel gauzes which are shaped in round cylinders approximately one and one-half inches in diameter and five inches high, which sit on top of the round safety glass. It is a regulation and a safety factor to have the second gauze to cover the initial top gauze. There is another gauze in the base of the lamp on top of the oil receptacle which has the important functions of ventilating the flame and moving the burned fuel elements into the top gauzes.

Despite the skill that experienced mine examiners use in gassy atmospheres during an examination of the environment containing ten percent of firedamp, methane's top explosive force, it is possible for the lamp to be "pulled" by the deputy in a slight motion toward his body. The second gauze inside the casing is the examiner's "safety factor", and would not let the flame burst into the open atmosphere. A jarring jerk inside the lamp when it explodes gas in the gauzes could force the flame through a single-gauzed lamp. However, extreme caution is the general practice for competent miners using the safety lamps in gassy underground environments.

The experienced mine official will detect firedamp in the mine air, generally, without exploding it inside the locked miner's light, but no attempt should be made to blow the burning gas out. It was a common practice for the coal face miners, during the days the flame safety lamps were used for lighting and gas detectors, to smother the gas burning inside the lamp under their clothing after a gas test. The flame safety lamp should not be allowed to linger and burn in an atmosphere which contains methane gas, and adding to this caution, the safety light which was exploded in fiery methane several times should be returned to the surface for another examination of the gauzes. Both lamps, flame and electric, are magnetically locked and only opened in the colliery lamphouse on the surface. In the large and more prosperous coal mines the underground people are supplied with portable electric lamps, electric gas testers, and the flame safety lights. Some coal companies have gas masks given out each shift to miners who work in gassy coal seams, but there are miners working in gassy mines who will never be given gas masks to save their lives.

The flame safety lamps are cleaned and examined after every shift, as the tiny flame of fire fills the open mesh of the steel gauzes with particles of black soot which would affect the security of the lamp in a gassy underground workings. Feed air for the lights combustion and its release from the casings makes it necessary for the gauzes to be rid of any substances that could obstruct ventilation into the flame. Most mining laws "demand" that each lamp taken underground will be scrutinized by competent people, other then the individuals who are to use the lights in the mines. However, the responsible mine official will be present when the flame safety lamp that he carries underground is assembled in the lamp house.

The mine boss, who has a flame safety lamp in his possession for the purpose of examining the underground atmosphere, should know before his light is placed in a flammable environment that it will be safe in explosive methane. The "wide-a-wake" mine deputy will have watchful eyes on each part of his flame lamp as it is assembled in the colliery lamp house. Then the competent mine "fire-boss" will "walk tall" in the mine, knowing that he has security for his fellow workers in his hands. However, although the people who care for mine lamps on the surface are competent and qualified, the practice of hanging flame safety lamps outside the lamp houses to be picked up by mine examiners and "shotlighters" should be stopped at some of the American mines.

The modern-day flame safety lamp will withstand the most abusive practices in the hands of unconcerned persons, yet precautions are necessary to gassy environments. This writer listened to a mine safety lecture given by an American mine inspector, who told that he took a safety lamp out of a mine that he found burning unattended in flammable gas with the gauzes burned and blackened with soot. In spite of all the security which the safety lamp affords mine "novices" in gassy atmospheres, men who should have known better have taken "leaky lamps" underground, through careless inspection when the lamp was assembled. Slipshod acts of assembling lamps and burning the gauzes is shameful negligence for any mine official.

Mine management may have reason to be concerned when mine bosses are allowed to use two lamps (flame and electric) in gassy coal seams. In some mining districts the underground supervisors are only allowed to use the flame safety lamp. Should the mine personnel be allowed to have both lamps? The supervisor who is not proficient in a gassy environment to carry the flame safety lamp and electric safety light as one piece is not a competent mine boss. The most important person in the mine workings is the man who has charge of the section, and he is at a disadvantage with only one lamp, flame or electric. The flame lamp will not illuminate in methane gas, although the firedamp in the air may not be poisonous. Does the "fire-boss" who never attended a colliery mining class know that his people could be working in rooms where the oxygen is so low in the atmosphere that he could not light a flame lamp in the environment? Electric lamps are not completely safe in explosive firedamp and should be used sparingly in a body of gas.

The first electric lamps used in the coal pits were unsafe in flammable air, and it was not until Thomas Edison produced his portable mine lamps that the Bureau of Mines, in the United States, approved the use of electric lamps in the workings of coal mines. It is time for top mine management to recognize that their underground officials are not properly equipped with only one lamp. The leading mine hazards are falling rocks and the underground transportation of coal to the pit-heads, however, both of these risks were reduced by the illumination of electric lamps.

The first English mine examiners were known as the "firemen" who searched the pits at night with open flames of fire, and it was not uncommon for them to burn small amounts of gas that accumulated in a room-end to clear the air for the miners who would work there in the morning. During the very first days of coal mining the firemen scribbled notes on a board at the mouth of the mine, which generally read, "all clear underground, or no admittance, the fireman is working." The fireman's dress for his job consisted of mostly leather and on top of that he had a long loose wool garment soaked in water before attempting to set any gas afire. History has written that the early firemen were often victims of fires and blasts during their efforts to clear the air in a "heading".

The first firemen made their own gas detectors on the "pit-bottom", before they started to examine the underground workings. At first the old-time examiner fashioned a ball out of the soft mine bottom clay that he formed in the palms of his hands, and then he cut the top off a small tallow candle that he worked into the moistened clay ball. This was the first mine gas detector of any significance. The fireman walked about the underground workings and when he suspected firedamp was in the surrounding air he backed up and lighted his clay lamp, and he lowered the flame to a tiny spark by moulding the damp clay around the burning candle top inside the ball. The examiner placed both hands around the open flame inside the clay casing and gently lifted the light aloft, close to the mine ceiling. The instant a gas cap showed on top of the flame, the fireman removed the lamp from the environment before it exploded the atmosphere.

Performances by the mine examiners now are similar to the functions that the first miners applied during their search for gas in the underground workings of coal mines. However, today the trained mine deputy will examine the atmosphere with a closed safety lamp, and he can adjust the flame and relight the wick with a safety device on the bottom of the light's oil receptacle. The element of risk is nil when the flame safety lamps are carried by skilled mine deputies. The mine examiner, who reads the percentage of gas in the atmosphere with a lighted flame inside a closed vessel, cares for his lamps, flame and electric, but this writer met people in the pits who abused them. In every underground coal mine the probability of locating gas in the atmosphere is always present, but the mine manager who does his job with people who were trained in the colliery mining classes never has the fear that someday his colliery is going to be blasted. Experienced mining people have safely moved thousands of cubic feet of firedamp out of the underground workings, but they were the only miners allowed in the gassy environments until the air was cleared and moving at less than one percent of methane into return airways. In another chapter of this book the author will describe the safety of adjustments of burning wicks in lamps during gas examinations.

The nights of burning gas intentionally in the mines have long passed into tales that are now laughed at when these stories are told. Nonetheless, it was a long hard road to rid the mines of practices that were endangering the workers' lives, mostly because coal face miners lost wages and returned home if their places were not available for work in the morning. Through the persistence of concerned miners, protective legislation has eliminated many of the mining risks that old miners put up with at work. Nonetheless, even the present-time workers in the pits are exposed, often needlessly, to hazards in

"A WESTERN CANADA COAL MINE"

"Cape Breton miners, unemployed, refused to work in several of Alberta's coal pits.

"And people have asked why!"

HERRIOT/CALGARY HERALD

the workings, and this should be corrected by the mine safety officials during their inspection of high accident rate conveyor transportation systems. The science of mining is improving the production of coal, but safety in the mines has the same problems year after year, and miners are losing "prime time" from their jobs by the lack of instruction at work that will reduce injuries and increase pay cheques. Many of the old hazards linger in the mines and have become "habits" with new workers and the colliery personnel. Mine safety engineers must become fearless in their duties as a "safety inspector", and not have government men to show them unsafe places in the underground workings. The necessity for an overall examination of the new transportation methods on dusty roadways is important now, before they become potential fire hazards and prone to accidents. The likely ignition of dust around moving machinery where coal is transported through narrow passageways can be eliminated. But somebody has to be the "boss" and tell others to get the job done, not waiting for a fire or a serious happening to close the mine for several months.

Comprehensible safety literature should be distributed to the colliers, but safety must go farther than the "mine mouth" and underground supervisors are obligated to instruct and discipline people on the job, "with tolerance". There will be a "rush" for coal in the near future, and the mines must not drift back into the days that George Orwell has written about in his classic book "The Road to Wigan Pier". Unless the miners take an interest in the mining schools, slavery will be in the new mines, not the workers' safety.

The first mines were ventilated by natural ventilation, the surface atmosphere forced into mine shafts by the weight of the atmospheric pressure. The functions of the atmosphere and the temperature of the air above the earth's crust also worked underground to supply the oxygen needed in the atmosphere for people working at coal faces. The production of an air current under the ground without any form of a ventilator is "natural ventilation", and that method of ventilating mines was used in the first coal pits.

The natural air which falls down mine shafts will increase in quantity and velocity when the temperature underground is warmer than the cold air which was forced into "intake-air-shafts". The two columns of air underground, warm and cold, will follow the rule of least resistance, and when the opening in one or two mine shafts has the least resistance inside, the spent air will flow through that opening, generally known as the "return-air-shaft".

The natural air underground is subjected to the heat of the

ILLUSTRATION

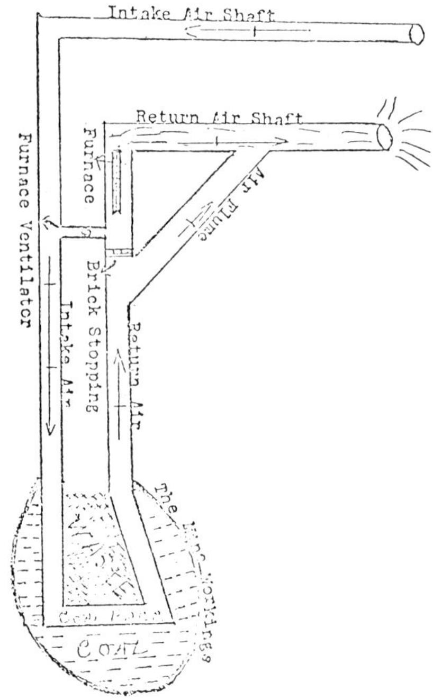

"THE LAST CENTURY FURNACE"

rock strata in the inner stratum as cold air expands at a greater rate in a warm atmosphere, and that dilation increases the flow of air into mine airways. The used air passing through the mine workings and out of pit air shafts was known as the "mine draught", and the English miners who arrived in Canada during 1900 used that expression in the mines. However, the natural ventilation was affected by the weather on the surface during seasonal storms, and it was stated that the first Nova Scotia miners in Pictou County watched the weather cock for favorable working days in gassy mines. There are many improvements in mine ventilation equipment all over the world, but gas is burning in the coal pits because mine bosses never attended a colliery mining school.

Ventilation in the coal mines was improved after the brick furnaces were installed underground, generally at the bottom of "return air shafts". The fireplaces increased the temperature of the air moving out from the mine workings which also boosted the quantity of air in the "intake air shaft". However, the pressure produced by the furnace will vary to the depth of the shaft where the heater is located. The only difference between natural and furnace ventilating in a coal mine is that the natural air underground is heated into movements by the temperature inside the earth, while the furnace fire will increase that temperature and the velocity of the air currents in the workings. During the natural ventilation of mines, and "if" the air in both shafts is of equal density, there can be no natural ventilating pressure, but with the furnace fire in one shaft, the dense mass of air is broken, then air will move in and out of both mine shafts. During the early days of mining along the banks of the Ohio River, in the United States, the furnaces were used in the pits, and smoke poured from holes in the ground, and visitors on the riverboats were told the mines had exploded. The "well fired" and constructed furnaces in the British coal mines were producing over 100,000 cubic feet of air each minute of the day into their mines.

For many years air from the furnaces was wasted on "leaky airways". The underground air passageways were not constructed with steel and concrete like many mines today. The first underground airways were just "worked-out-rooms", which were partly re-timbered, "falls", large amounts of stone which fell from the mine roof blocked-off the ventilating currents of air, were not wholly removed to allow the full quantity of air to pass into the mine workings. The air lost from these old airways mixed with gas in worked-out mining sections, old rooms, and wastes back of longwall coal faces; later, the gas and air moved at flammable percentages on the top of "wall workings",

"A." A mine shaft fan, which was placed on top of the "pit-ari-shaft."

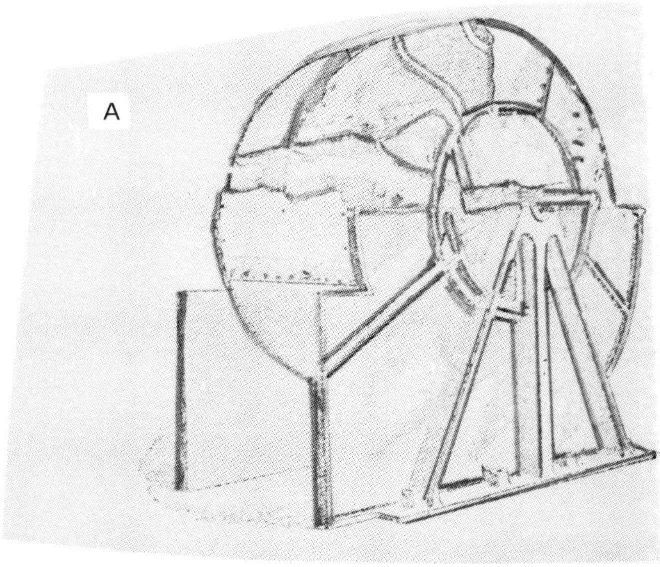

"B." A moveable underground "booster fan," generally placed in the underground workings during development of rooms and longwall sections.

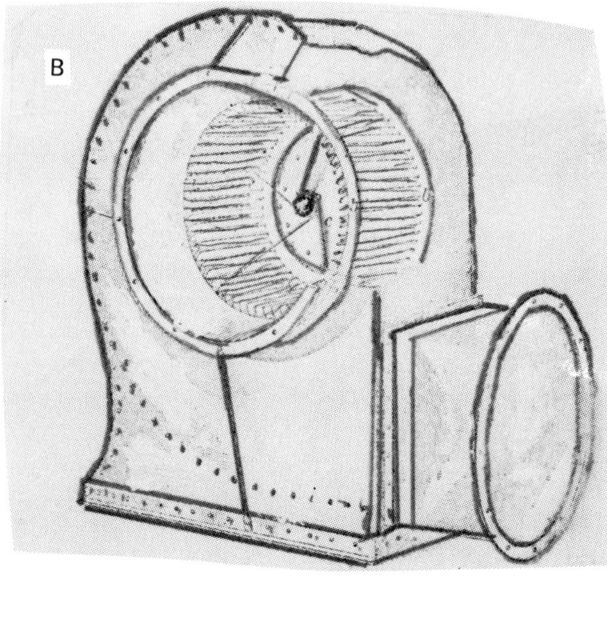

and at times into return airways used as haulage roads. The term used in mines for gas found on longwall material roads is "gob gas" and it is poisonous.

Before the centrifugal fans were introduced at the collieries during the last century, the furnaces were forcing high velocities of air into under-sized mine roadways and much of that air never reached the coal faces. The velocities of air in some of Britain's mines was moving so fast from furnace power it conceived the idea that Davy's gauzed lamps had to be protected in the mine airways, and a tin case was placed around the great scientist's lamp.

The centrifugal fans were not the first mechanical ventilators at the colliers; "air pumps" were used for several years to ventilate mines. The air pumps displaced air underground by pumping used air out of the mine, which left a vacuum in the underground atmosphere. The vacuum left in the mine atmosphere was immediately filled with new air which was forced underground by the world's atmospheric pressure on the surface of the earth. The "pumps" were known as "displacement machines", operated underground from steam piped from boilers on the surface.

The centrifugal fans used now on the top of mine shafts and underground can function as force or exhaust ventilators to clear the mine workings of noxious gases. The force ventilator on the surface will create an energy above the atmospheric pressure inside of its air shaft, and the exhaust fan which pulls the air through the mine workings will create a pressure force below the atmospheric pressure on its return air shaft. Most of the information on mine ventilating problems was taught to this writer in the Glace Bay mining schools by Bernard Currie, a colliery mining engineer. However, many of today's mining students forget that the atmospheric pressure remains an important part in deep mining projects.

The essential part of ventilating large mines is to have ventilators on top of mine shafts which are powerful enough to sweep all harmful gases underground, under normal conditions, out of the pits below flammable percentages. But at the same time the collieries have to develop "pit room" below ground, and any advancement of the mine workings into the earth's crust will create a greater "work load" on the surface ventilator. It isn't always practical to construct air shafts to reduce the fan's work, as mines are often extended under water and other restricted regions.

Fluids move from a point of greater pressure towards the point of least resistance underground and on the earth's surface, and that is the principle which a production of air depends in

"A" The colliery "fan-house" containing the enclosed mine ventilator.
"B" The mine "water-gauges" were generally placed in the fan-house on surface, but they were also placed underground.

A

B

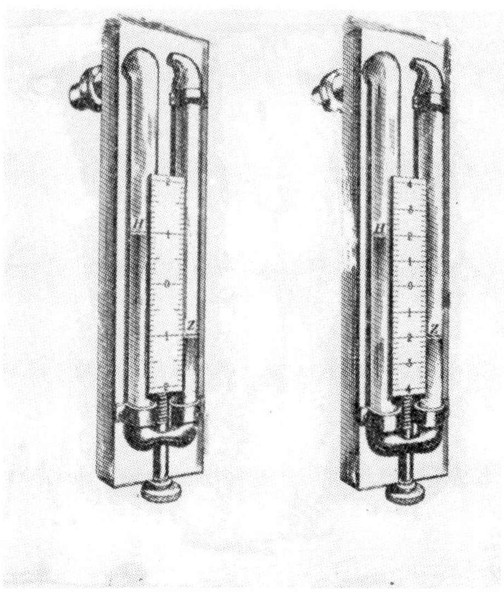

deep coal mines. Underground airways offer a resistance for the full passage of air generated by the mine ventilators, and this problem is known to mine students as "the rubbing surface resistance".

The length of an airway and the obstacles which the column of air contends with underground will create a greater work load on the fan, and reduces the quantity of air which the ventilator is capable of producing. The resistance which a mine airway reduces the volume of air produced by the mine ventilator is measured by a "water gauge".

The water gauge is placed in a permanent position in a wall at the fan house, or in a mine stopping built between two columns of air at different pressures. One arm of the open glass tube has an access to the greater pressure of the mine's intake air, while the other open end of the U-shaped glass remains in the lesser air pressure of the air returning out of the mine. The water is levelled on both sides of the glass instrument, but the higher pressure on the intake air will depress the water on its side of the glass; and the water will rise in the tube on the other side. The length of the water rise in the glass is measured in inches, and will give the approximate resistance the underground airways are reducing the ventilator's production of air. This is known in the mines as the "water column". However, the water gauge measures the resistance of air, not the volume of an air current moving into the coal face. Nonetheless, the "work-load" will be reduced on the ventilator and the quantity of air increased underground by removing many useless objects that are obstructing the movements of the ventilation in and out of the coal pits. The water gauges were improved over the years and it may be possible, during emergencies at the collieries, to record "mine bumps", and minor blasts of gas in a sealed mine with the water gauge.

In large mines that are worked deep in the earth's crust, permissible ventilators are placed in permanent concrete buildings underground. Although this practice was frowned upon at one time by mining men, it is necessary, and will also be used in the new mines which will be operating under the coastline of Cape Breton. The large mine ventilators will reduce the work of the surface fan and increase the velocity of the air underground. Close inspection is recommended at mines that operate with fans above and below ground when the underground ventilators are down for repairs. The water gauge will have a much higher reading in the surface fan house, that tells the mine officials that the pressure is reduced on the coal faces and the underground wastes. It is a known fact that during the days of natural ventilation in the mines, the gases moved out of the "old-workings" when the atmospheric pressure was reduced on the earth's surface, and it will happen now in mines where permanent fans are needed underground. The water gauge on the surface will return to its normal reading after both fans are in operation.

Small electric "booster fans" often placed in the immediate mine workings will become a hazard underground when they are not maintained and operated by competent people. The booster fans are generally used in development work in the pits, and the section "shotfirer" usually has charge of the operation. There are several dangerous practices which are connected with small fans close to the coal faces; forgetting to shut down the fans between shifts, and shutting them down on the weekends. The long lay-off during the week and the next start in headings which are kept clear of firedamp with small booster fans will load up with explosive methane gas. It is a general practice to turn the full force of the fan into this gassy atmosphere, although one realizes a tiny spark of fire would blast this gassy air.

Fig. 508. - Spedding's Steel Mill.

The most notable lighting appliances in mining history before the safety lamps were introduced in England during 1815, were the "Steel Mills", and they were used in coal pits for over fifty years. The steel mills were used for lighting in the English mines from 1760 until 1815. Repeated fires and blasts at the collieries made it necessary to have underground illumination which would not fire flammable mine atmospheres, and the miners and the operators for twenty-five years claimed that the "mills" would not set gas afire. Mine explosions were confined in the early coal pits, and its effects were not widely circulated throughout the workings from the lack of oxygen, but smoke and poisonous gases trapped miners in other sections of the mine.

During the 1800s, and parts of the last two decades of the 1700s, underground furnaces were pulling thousands of cubic feet of air into the mines and the underground explosions became more severe. The sketch of the steel mill shown here was taken from a collier's handbook which was published in the last century. The steel mill when in use was spreading sparks of fire into the mine atmosphere when the operator pressed a hard piece of flint quartz against the front wheel and at the same time turned the small crank which was attached to the side of the mill. Each mill was "played" by one or two strong "pit boys", who moved it about the mine when it was necessary to have illumination away from the coal faces. There were people who became suspicious of the mills not being capable to fire gas in the mines, but they did not wish to restrict the only lighting apparatus that the colliers "swore by".

The trust in the steel mills and the colliers' faith in them was shattered after 25 years, when a survivor of a blast told that the mill he was operating exploded the gassy atmosphere. Several more fires and explosions and the miners realized the mills were the culprits. Nonetheless, the mills played on for another quarter of a century, mostly because they had no other form of mine lighting except the open flame of a candle or a torch. It is possible that the steel mills did not have the same capability to heat low percentages of unexplosive methane to an ignition temperature as the open flame of a candle. The tiny sparks of fire produced by the mills instantaneously blacked-out in low percentages of oxygen in immediate atmosphere, but it later became known that the lack of oxygen around the coal faces rendered the firedamp unflammable in the early coal pits.

Fifteen years before the safety lamps were invented, and used in the English coal mines, a safety lamp was experimented with in the German mines. Friedrich Von Humboldt, a young mining student who became a noted explorer and scientist, was constructing a mine lamp late in the 1700s. Humboldt, as a young man, was a very capable official in the German coal industry, but he left the mines after a dispute and the lamp he was planning did not precede the mine safety lamp in England. German writers have written, "Humboldt invented the mine safety lamp before 1815," and others have stated, "that might have happened if the German scientist had remained several more years in the coal industry." Although Humboldt is known as an explorer, a scientist and an author in history books, he attempted to cover all of his scientific knowledge in a volume of books named "Kosmos". The German scientist was writing those books when he passed away at 90 years of age.

W.R. Clanny, a coal town medical doctor, was teaching miners the functions of first aid to the injured early in the last century. Although colliers have gone into the pits during fires and after explosions since the first days of mining, it was Doctor Clanny's pupils who conditioned and skilled the primitive "mine rescue people". The first mine rescue corps at the collieries went underground during emergencies, in smoke and fire, "barefaced" to rescue comrades, and heroic acts are attributed to their bravery over the years. The early decorations for miners who often sacrificed themselves in fiery mines to save others was the Saint George's medal. The underground fire fighters nowadays are the same breed of colliers, miners, mine bosses, and at times colliery office clerks; all are healthy and intelligent people, trained underground for unforeseen occurrences.

Most of the mine rescue people now working in the mining districts hold "The St. John Ambulance Association Certificate", in first aid. The rescue trainees, after they become proficient in handling the appliances needed in a fired coal mine, are grouped into teams under one leader. That person will follow instructions planned before they left the surface while his squad is underground. Nonetheless, it may become necessary underground to change planned directions, then he will let his team know of these changes without going into any long lecture, as time is important when fires are burning in the mine workings. The mine rescue men have now another safety device for protection and instructions underground — the transistor portable walk and talk radio-telephones. The radio-telephones can inform others where the fire fighters are located and the condition of the fired workings.

Doctor Clanny is not remembered in mining for the doctrinal he taught the colliers, but a lamp that he showed at a first aid class during 1813 had his name on it in mines for over 100 years. The Doctor's first "clanny lamp" was isolated from the open atmosphere by water inside its casing, and two small bellows which were attached on the outside of the vessel were used to pump air into the flame. The initial Clanny lamp blacked-out in a gassy atmosphere, and was of little use to the coal face workers; however, at a later date the doctor improved his lamp with other people's inventions, and the Clanny lamps were considered the most improved safety lamps for the underground workers for many years. The Davy and the Clanny pit lamps were similar in their construction and both used the same principle for lighting and examining the mine air. During the years from 1835 until 1840, there was a suspicion at the collieries which concerned the safety of the flame closed lamps in gassy underground environments, and Doctor Clanny, then an old man, defended his lamp before an English Mining Institute during the 1840s.

George Stephenson was the son of a mine fireman, and at an early age became his father's assistant. George's first school days were neglected like many other "little children" working in mines and factories at the age of 10 years. Although Stephenson lacked an early school training, he was a studious boy who gained an education from books that his father purchased for him to study at home, and he was a mine manager during 1815. This mine manager was involved in another invention during 1814 — the "steam blast" which is used in locomotion, and which the miner used to pull coal out of his mine. Stephenson was aware that power-driven machinery would become necessary in the mines, so he built an engine which hauled up a steep colliery slope 30 tons of coal at four miles an hour. The manager's engine was a firstling, and railroad historians have written, "George Stephenson started all of it in his colliery yard during the year of 1814."

The colliery manager who became famous as the inventor of the "steam blast", constructed locomotives and railroads, built iron bridges across wide rivers, and later became a member of the British House of Commons, is not recognized in written history as the person who invented the miners' flame safety lamps during 1815. Nonetheless, George Stephenson was the only person to build and use a practical flame safety lamp in a mine during the year of 1815. The sequence of events to be read in the following paragraphs may leave the readers of this book sometimes confused in regard to which person invented the flame safety lamps, Davy or Stephenson.

The mine manager was experimenting on the combustion in closed lamps at the Killingworth mine, on the River Tyne, during 1813. Although there were a few mining engineers who supported Stephenson's claim, they knew the manager had the most advanced information on the closed lights before Davy had any notion of going into the colliery districts and searching for a lamp that would save miners' lives. However, many people supported the great scientist, and said, "Stephenson was not a chemist." He did not have the scientific knowledge which induced the findings that caused the invention of the "flame safety lamps". The strange part of this was the division of the miners to support Stephenson, after he claimed the lamp's invention.

The Royal Society of London at one time asked the British Parliament to contain deep mining until safer methods were developed to remove the coal from their mines. Britain became a great power and the commodity of coal became its greatest commercial export. But, the coal trade would have been delayed if the mine safety lamps were not invented for the coal miners during 1815. "Sir Humphrey Davy" was not interested in the coal industry at that time, but a gentleman named Banks, who was president of the Royal Society during the year the safety lamps were introduced in England, coaxed the scientist (Davy), to visit the mining districts where underground explosions were becoming almost a daily occurrence. Mine fatalities were increasing at the collieries to a critical point, and people in high places became concerned with the miners' welfare in the pits and the industry.

Davy made it to the coal pits along the River Tyne in Newcastle with a young man named Faraday, who became his assistant during gas experiments at the collieries. Davy and Faraday gave ten years of scientific study to the coal industry through which mining was elevated into a knowledgeable study for the people who inhabited the mining districts. Faraday's mathematical formulas are used in mine ventilating problems in the colliery mining classes, and his "coefficient of friction formula" is one unit of resistance on a mine airway.

Michael Faraday was a bookbinder who demonstrated to Davy his skill in mathematics, and with his association with Humphry he became a noted professor in the "laws of electrical science". During later years Faraday was knighted, and also became one of the presidents of London's Royal Society for his achievements as a scientist. This writer did not find any history that connected Faraday with the safety lamps at that period, although it is over 150 years since Professor Faraday warned the coal operators to beware of the coal dust which floats in the mine air. For many years after that statement, and numerous mine explosions in every country, chemists were wondering about what Faraday had written and what he meant; as many of them have stated, "coal dust was not explosive in itself."

Late in the last century two noted French chemists, Mallard

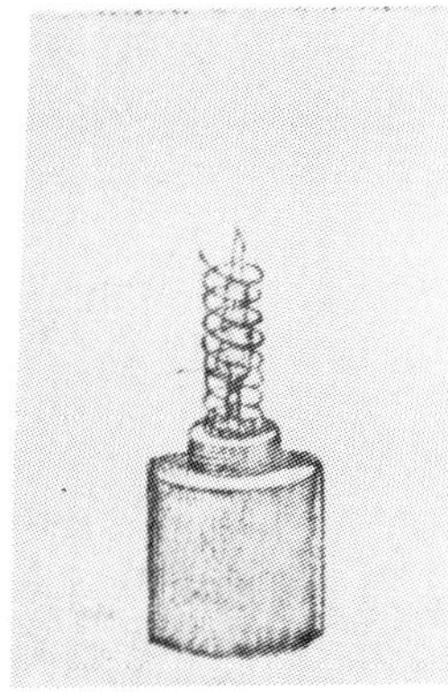

Fig. 509 - Experiment with Alcohol Lamp.

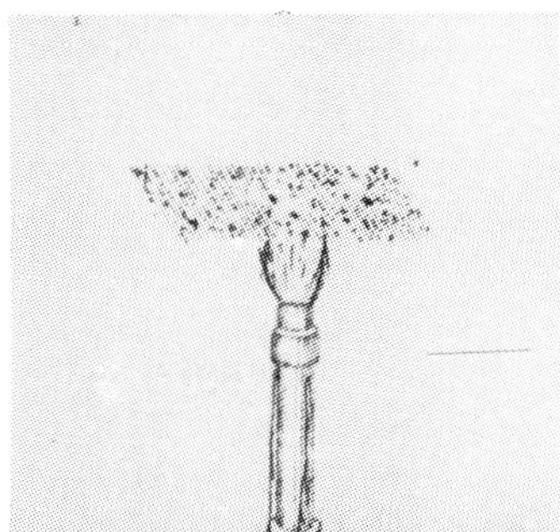

Fig. 510 - Experiment with Flame and Iron-Wire Gauze.

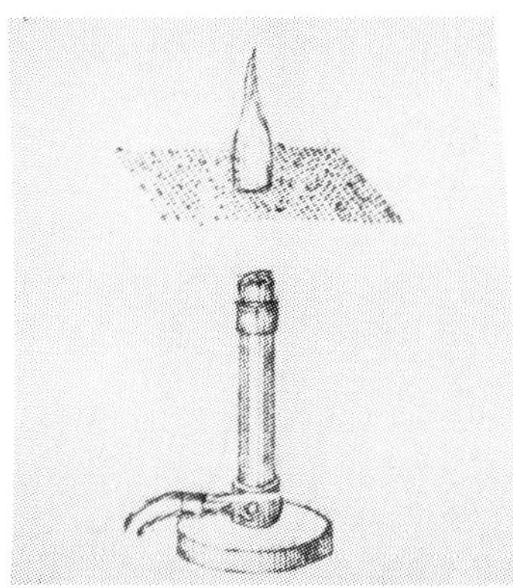

Fig. 511 - Experiment with Flame and Iron-Wire Gauze.

and Chatelier, reported "coal dust was not a danger and no colliery explosion could be derived from coal dust." Thousands of experiments were made with coal dust, and many of them in this century, before most mining people realized coal dust was "explosive in itself".

Sir Fredrick Abel, an English chemist, was considered the outstanding authority on explosives during his lifetime, yet he died during this century believing that something besides coal dust had to be in the mine atmosphere before it exploded. Nonetheless, Abel did recognize the danger of coal dust in the mine atmosphere when he told coal dust skeptics that less than two percent of unexplosive methane gas mixed with coal dust was a potential explosive atmosphere. Later, that same chemist (Abel), told a group of mining engineers that the same dust he used in his experiments need not be flammable to produce the results that he found in the experiments which exploded when the mine air contained small percentages of firedamp. That message from Abel should be a reminder for mining people nowadays who are told coal dust in many mines is not explosive. There is no way that mine dust can be anything but explosive during continued mining operations on "longwalls" as each cut gives off percentages of methane gas. Generally, rock dust and water are used to reduce the flammable dust in the mines, but mining people will have to find some other means to combat fiery coal dust that the colliers are breathing in many pits now. Dust samples taken from well-kept mines will have greater percentages of non-combustible materials on the floors, sides and ceilings. It is a known fact that the fire from the blast in Number 20 mine at Glace Bay was stopped by the amount of stone dust in the airway leading into the coal faces.

Sir Humphry Davy was not of peerage birth. He was the son of a wood carver, who apprenticed him to a small medical school at an early age. Davy became a brilliant student in medicine, but turned his talents towards chemistry. The great scientist was a fine poet, and he might have become one of the distinguished bards of his day, only his love for science was greater. Davy wondered from one set of people to another, equally at home in any society. Humphry discovered early in his career the sting of critics, and he was scolded for a paper he published under the name "Chemical Science", when he was in his teens. Davy as a young chemist proved that nitrous oxide could be used in hospitals as an anaesthetic, but the method he used in this finding was not considered a practical means for a chemistry experiment. Sir Humphry was treated as a "nonchalant fellow" by other chemists at that period, but he became a skillful physicist who will never be forgotten in our world. The great scientist was more than practical when he formulated an electrical theory of chemical affinity, which was verified as accurate 100 years after Davy's death. He was only 51 years of age when he died of an illness contracted during chemical experiments.

George Stephenson was the manager of the Killingworth Colliery on the River Tyne, near Newcastle, during 1815, and in August of that year he took a brief "time off" from his mine. Several days later Stephenson returned to his duties at the colliery, and he immediately summoned the mine surveyor to sketch an outline for a lamp that he would describe. The mine manager completed two lamps before the Fall of 1815, which burned safely at a "gas blower" in the Killingworth mine. The second lamp made by Stephenson had improved burning power inside the casing; it also was tested in the mine gas.

Several weeks after producing two mine safety lamps, Stephenson showed his third light and claimed the invention of the miners' flame safety lamps. However, it should be stated here, the mine manager's lamps were gas detectors, and the only illumination which they produced was a glimmer of light from inside the closed casing.

Both Davy and Stephenson showed several forms of mine lamps during 1816 and 1817, but people were confused on the "principles" of the lamps which were not fully explained to the colliers or the public. Men and women were debating the merits of each lamp; they did not have the information on why the flame inside the casing, which was drawing feed air from the mine atmosphere, would not explode flammable methane outside the lamp's casing. Fires kindled from any material are burning gases; shut the oxvgen off from the flame and the fire will die. People were trying for illumination in gassy underground atmospheres with closed lights which had little or no ventilation, and when they placed air vents in the casing the burning wicks heated flammable firedamp on the lamp's exterior into a fire or a blast.

Sir Humphry Davy visited the colliery districts surrounding Newcastle, on the River Tyne, during the month of August 1815; maybe George Stephenson was among the people who greeted the scientist along the river banks that sunny August day, but this was not told by Davy nor the mine manager. Is it possible that the scientist Davy never met the man who invented the "steam-blast" for locomotion until the year the mine safety lamps were invented? But that is the way it is told in mining history. However, it is known that Stephenson and the scientist did meet around that time, although it is only assumed they discussed mine safety lamps. Davy was around the mining districts from August until November, during 1815, before he returned to the City of London. During the scientist's short stay in London, before returning to the mining districts he read a paper containing fascinating accounts of his experiences with the gases found in coal mines. Davy told his audience that methane gas had to be heated into a point of ignition before it burned or exploded, and that "cold surfaces" retarded heat and would actually put out a flame. The results of these findings by Davy were the "principle" in which mine safety lamps were invented, and used today in gassy mine atmospheres.

The scientist was experimenting with fire, iron and methane, when he noticed the flame, which was placed directly under an iron screen, was not penetrating the gauzed metal — not until the cold iron mesh was heated into an ignition temperature. It was through this experiment that Davy solved the mystifying efforts of others who thought about safety lamps for miners, but never accomplished "a flame lamp which would be ventilated from the open atmosphere in gassy coal seams."

It was late in the month of December, during 1815, before Davy placed and tightly secured a small lighted wax candle inside a closed circular iron gauze, and that "candlestick" was the first mine safety lamp made by the great chemist. The burning candle surrounded by the wire screen was placed in a large wooden box which was later filled with explosive methane gas. The firedamp was exploded in the box, with only a scant amount of gas seeping into the iron gauze. The flammable air and methane on the outside of the lamp's casing remained stagnant inside the gassy box. The "cold surface" of the metal gauze on its exterior kept the flammable gas surrounding the scientist's light below an ignition temperature, and could not be fired. However, the explosive firedamp inside the iron gauze lamp exploded without any impairment to the wooden box. Generally, the coolness of the iron sleeve inside the light will black-out the flame after the gas explodes inside the gauze during a gas test, but it may burn in the lamp for an instant until the examiner removes the lamp into clear air.

Davy proved the mine safety lamps were safe in gassy atmospheres with the experiment in the wooden box, but he was not ready to allow his invention to be taken into a mine workings, and he waited until the new year before it was used in a coal mine. The public was clamoring for information on the new lamp, but the "principle" was never fully explained to the public at that time. Although Davy had no secrets about his invention, he disclosed the mystery of the flame safety lamps that miners are using now in gassy coal mines, in a very seclud-ed paper, published by the Royal Society during 1816.

"ILLUSTRATIONS"

DAVY'S FIRST LAMPS – "A" - THE CANDLESTICK

DAVY'S FIRST OIL LAMP – "B" - 1816

DAVY'S SAFETY LAMP – "C" - 1817

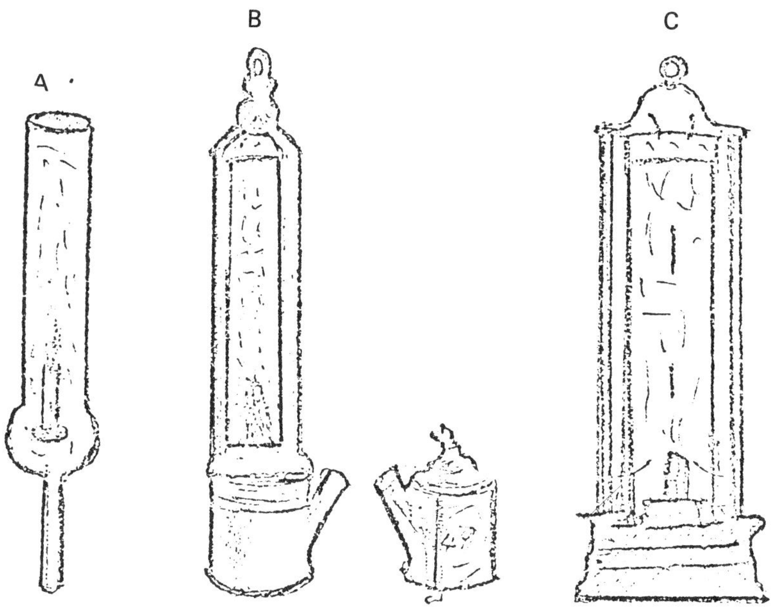

The peculiar and disturbing substance of Davy's message on the hazards and safety of the lamps he invented was never explained to the majority of people who have used flame safety lamps in the mines since 1816. In spite of history books which tell that Davy invented the safety lamp which saved many miners' lives, it is no secret that Stephenson was using safety lamps underground during 1815, although some mystery was connected with the principle of the lamps and how the inventor acquired the information to build a safe mine lamp.

That same "principle" that Stephenson used in his first three lamps is working in today's safety lamps during gassy examinations of air in the coal mines. The fundamental procedure that Stephenson used in his lamps was an effective addition to smothering the fire after methane gas explodes inside the modern-day flame safety lamp. For many years most of the mining personnel clung to the conception that the metal bonnets which cover the flame safety lamps were only there to protect the gauzes from dust and high velocities of air. The lamp's bonnet which protects the steel gauzes from the "wear and tear" of mine work is also extinguishing gas which burns inside the safety lamp during an underground environment examination. Nonetheless, very few underground workers were ever told this; many years ago, when I was a young man in the pits, I asked an underground official to explain the safety lamp to me. He answered, "All I know is that it works in gassy atmospheres." The burned elements from the fuel flame in the safety lamp will not support combustion during an examination in gassy mine air. The burned substances from the fuel flame were intentionally restricted by Stephenson to the inside of the casing which surmounted his lamp which did not use the gauzes that Davy had in his lamps. And that was the principal difference in the lamps invented by Stephenson and Davy.

The average miner, searching for gas in the mine workings with a flame safety lamp, may have a notion to lower the fuel wick before placing the light in a suspected body of firedamp. The initial probe with a safety lamp into a gassy environment should be made with the normal fuel flame, "not the lowered wick". The combustion from a medium-size flame in a burning safety lamp nearly fills the interior of the bonneted gauzed lighting device with carbon dioxide, a non-combustible gas. The trained mine examiner will remove himself from the edge of a body of explosive methane and re-light his lamp; then he will lower the flame and examine the atmosphere with caution for the percentage of gas in the immediate environment.

The three lamps invented during 1815 were gas testers and

The "Tin-Can" Davy mine lamp after the "window" was placed in it. It could not illuminate the darkness, as the iron gauzes surrounded the flame in the lamp.

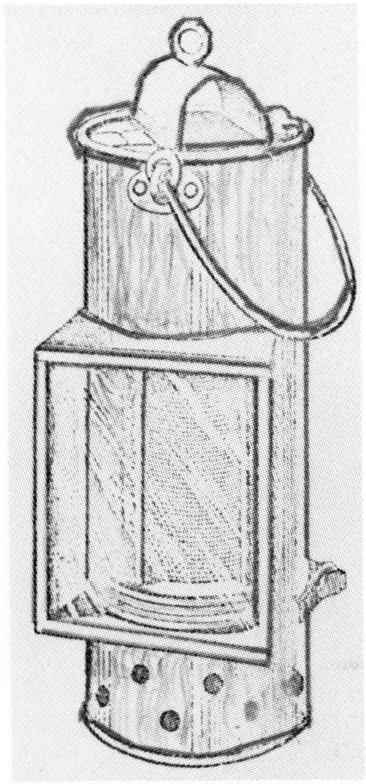

Fig. 515 – Davy Lamp In Case.

did not give the bearers any light in the mine darkness. Stephenson's lamps were the most popular at first, but Davy's lamp was considered the best light for use by mine examiners. The Clanny lamp became the safest lamp used by the "run-o-mine worker" in later years. There were other types of mine safety lamps manufactured after 1817, but most of them were copies of the lamps designed by Davy and Stephenson. The new safety lamps did not improve the miners' "eye sight" in the pits, and the miners refused to take them from the lamp house which caused strikes in several districts; then the government passed legislation and forced the English colliers to use safety lamps.

During the last two months of 1815, George Stephenson made several noted changes in his lamps which admitted air into the closed vessel. The most significant rearrangement at that time was the removal of the small copper tubes in the base of his lamps used to ventilate the flames. However, the mine manager made a major change in his lamps during 1818. The ventilating gauze placed under the wick of the modern-day lamp is an important installment and Stephenson was the first to recognize this when he placed tiny copper tubes in the bottom of his first two lamps. The ventilating air which supplies the lamp's lighting power through the base also has another significant role — "it creates a pressure in the bottom of the closed device and forces the burned combustion into the top gauzes and out of the lamp." During the next few months this writer will search through libraries, books and papers for information on Stephenson's third lamp for the exact details on how the mine manager ventilated the flame after he removed the miniature copper tubes from his first two lamps.

Davy may have hesitated or delayed underground experiments with his gauzed lamp for security reasons. It was 1816, on New Year's Day, at the old Hebburn Colliery in England, that a thoroughly instructed fireman carried the scientist's invention down into a coal mine. That old mine fireman examined a gassy mine workings with the first "Davy pit lamp", returned to the surface, and reported his findings in the mine log. Not long after, manufacturers besieged the scientist to name a price for the "gauzed lamp", and Davy replied, "I'll never take money for saving the lives of the coal miners." He never did.

The mine manager and the scientist were simultaneously building mine lamps during the same periods, although there is doubt that both men used each other's guidance. Davy told Stephenson, without any abashment, that the glass and copper he was using in his lamps would become a hazard underground during the long shifts which the lights burned in the mines. Sir Humphry was respected by the miners as a person who came to help them, and Davy became concerned for the safety of the people working in the coal pits and the colliers who lived in the mining communities. The scientist was an intellectual chemist whose first interest at the collieries was the mine gases, and through experiments with the gases the miners' flame safety lamps were invented.

Soon after, it was learned in the mining districts that Davy refused compensation for the gauzed safety lamp's invention; people in the coal trade along the River Tyne presented the scientist with a silver plaque worth more than $10,000. Davy's brother, a medical doctor, placed the plaque in the archives of an English institute in later years. Stephenson was also remembered by the same people along the Tyne River, and he was given a purse containing $5,000 for his contribution to safe lighting in gassy coal mines. Doctor Clanny, one of the three persons who showed mine lamps during 1815, improved his mine lamps during his lifetime, and for many years, in mining districts all over the world, all of the flame safety lamps used by the miners were known as "the Clannys". During the early decades of the 1800s, Dr. Clanny did more than others to update his lamps.

This sketch shows a very early "bonneted" mine lamp where Davy used the drilled circular iron ring to ventilate the flame inside the closed lamp. George Stephenson had used the drilled copper tube below the glass in the third lamp that he made during 1815.

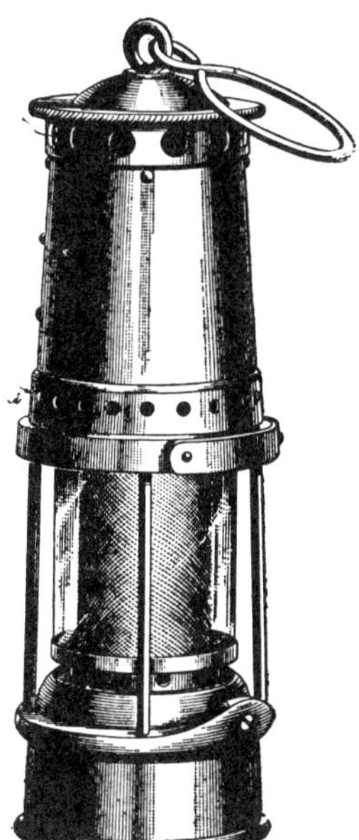

This lamp became known as the "Jack Davy" pit lamp.

Twenty-five years after the flame safety lamps were introduced in England, the closed lights had their first real setback. It was reported in the 1840s that more people were lost in the mines since the lamps were invented then in the days the steel mills were sparking fire in the underground workings. These statements may have been true, as the coal trade advanced from 1825, and many more people were in the mines during the periods mentioned in the last sentence. The coal industry made Britain's great sea power which controlled most of the world for nearly two centuries. There was no official records of mine fatalities at the collieries until 1850. During that year the English government sent mine inspectors into the pits to police the underground workings. Most of the early clergy in England and Scotland have kept records of mine explosions and the number of people lost in each blast, during their assignments in coal villages. An old Anglican priest has written into his parish records the following statement. "The blast occurred at five o'clock, A.M., the woman burried today in the church yard worked on the pit bottom, and when the explosion happened she was blown up the mine shaft and her bowels were removed from the pulley wheels." Although men and women worked underground in the English mines at one time, females were taken out of the coal pits early in the last century, but married women are now working in the United States' mines. However, I have read that women were in the Welsh mines during the first World War, and they have been underground in Russia since the first days of mining. Did women ever work in the Canadian coal mines? "No." They may have worked on the surface "pit heads", but never in Cape Breton.

There should be no doubt that Stephenson made the first mine safety lamp, nonetheless, it was the subject of bitter discussions in "pubs and castles" for many years. On November 30, 1815, the manager of the Killingworth mine carried his third lamp underground and boldly walked into a body of flammable methane and returned to the surface. The first lamp made by the manager had one copper tube in the base of the lamp to ventilate the wick; the second lamp had three capillary tubes to admit air into the flame. He removed the copper tubes from the third lamp that he made in 1815. None of the mine manager's first lamps used the iron gauzes inside their cases. The mine manager's lamp was known as "the Geordie light" around the collieries, and it weighed approximately two pounds; it had a glass chimney five inches high and the diameter of which was nearly two inches. The glass chimney was covered with a cap of perforated copper, which restricted the combustion of the fuel flame inside the chimney to black-out the fire during a gas test in fiery methane. The glass chimney sat on top of a ringed copper tube which had small holes drilled into it for passage of air into the flame, all of which was completely surmounted by a circular copper shield (a bonnet), which was also punched through with hundreds of tiny pin holes. Because of the numerous perforations in the Stephenson lamps, it compounded the principles of the scientist's and the mine manager's invention into one and the same, which was not correct, and that is what the writer is hoping to explain here. It must be remembered that information did not move with the speed of present communication, during the early 1800s, and it was a few years later before people learned the different designs of both lamps.

During 1816 and 1817 the safety lamps were improving for the mine examiners, although the underground workers were looking for illumination at the coal faces. However, it was in 1818 when Stephenson placed the iron gauzes inside of his lamps. The Davy lamps were going through changes during the same period, and the lamp with the tin case had a window placed in it. Several months later, a Davy lamp with a moveable glass cylinder on the outside of the gauze was constructed for shot-firers. The shot lighters lamps with the glass cylinders were also used by the mine examiners when other lamps were not available, as they were better than the average Davy lamp in gassy places. The glass cylinder was generally lifted sufficiently to allow the shot lighter to gently push a small wire between meshes in the gauzes to heat the wire around the burning wick and then the red-hot wire was used to light explosives at the coal faces.

During the first years after the gauzed lamps were introduced, they were regarded as safe mine lights, but mining methods changed and the ventilating air currents in the pits were moving faster, and it is a fact that people using Davy's name after his death in Switzerland during 1829, allowed the lamps to deteriorate. Manufacturers made profits in the construction of the early lamps, but improvements were slow to happen as the lights aged. Explosions continued in mines all over the world, and often the closed lamps were blamed for lighting the gas which exploded in the coal pits. The "run-o-mine" workers were using lamps which were carelessly assembled before given to them, and a few lamps were not locked. However, it is well to remember the safety lamps were made safe by men who worked and studied for the miners' safety underground.

During the last quarter of the 1800s, the British government appointed a noted chemist, Professor Galloway, as a mining inspector. The professor was given a special assignment under "Queen Victoria's" signature to find the cause of the numerous fire ignitions in their mines, which had caused many deaths in the colliery towns. The professor was a college chemist, and he searched through the underground darkness for factual evidence on the underground fires; generally there were no survivors after the blasts, nor did the workers give him much help. Galloway worked many hours each day, and he should be remembered by the mine inspectors who are now searching through gassy mines to save miners' lives. During the last century and for a few years in the 1900s, gunpowder was used in the mines for blasting; it flamed after it exploded at the coal faces. However, most of the English miners were colliers from birth, and they learned about pits and gas at an early age; they knew enough about mining to clear the air at the coal faces before firing a shot, anyway most of the fires and explosions were occurring away from the room-ends. The only other flame of fire during these periods underground besides the flame of blasting powder was the tiny light burning inside the safety lamps, and once more in the 19th century the lamps became a target of suspicion, and the mine inspectors forced a review of the safety lamps. During 1886, a British Royal Commission reported that if a mine atmosphere becomes gassy, the Davy and Clanny lamps, and in a lesser degree, Stephenson's lamp, are unsafe in currents of air having velocities much below those encountered in well-ventilated mines. The commission made no hesitation in stating, "Prohibit the lamps until they were made safe in gassy mines."

The British mine inspectors warned the safety lamp manufacturers to repair their lamps at once or they would be taken out of the mines. The safety lamps were renovated, but other people were building new gauzed lamps in Europe that did not use the first three inventors' names, and it was stated the new lamps were an improvement over the lamps in the English mines at that period. Several of the people sitting on the 1886 Royal Commission named four lamps that they said were deserving of special attention. They were Gray's lamp, Marsaut's lamp, the bonneted Mueseler lamp, and Thomas's modification of the Clanny lamp. The first inventors lamps were in such a condition that they would explode firedamp in currents of air with velocities of less than 800 feet a second, while the new lamps were able to resist in highly explosive atmospheres, with velocities of 3,000 feet a second Nonetheless, the old lamps prevailed, but they were allowed to deteriorate in the mines during the 1900s. These three lamps bearing the names of Stephenson, Davy and Clanny, were barred from the United States mines, but the mine deputies in Pennsylvania coal regions used the Davy lamps to examine the mine workings for gas. The flame lamps became a hazard in the grimy hands of a coal face miner, because the lights needed replacements, and through human carelessness underground workers were taking lamps into the coal pits which exploded in fiery methane. Although most of the credit for safe mining light belongs to Sir Humphry Davy, George Stephenson also has credit for the years he spent to build a safe lamp for his own people, and also after he used the gauzes. The first pair of expansion rings to hold the gauzes and the circular glass tight inside the bonnet was made at Stephenson's mine forge.

James W. Paul was head of the Bureau of Mines in the United States during 1919, when he prepared a written statement of nineteen points on mine safety; several of the sentences in the paper touched on the safe use of the closed flame lamps. Paul was talking and writing notes on safety in the pits during the years when mixed lights were used by the miners — open and closed flame lamps. The need for new people to be instructed before they become underground workers is much greater now than it was when "compressed air" was used to operate mining machines. Mr. Paul was the first secretary of "The Mine Inspectors' Institute of America", and was one of the Institute's organizers during 1907. During the period that Paul was a mine inspector, most of the mine workers were employed on contracts, piece work, and it was often stated "that it was impossible for miners to work in a safe manner, and earn a living wage."

For the past seventy years the American Institute of Mine Inspectors has spent time and often heartaches in the coal pits, hoping to improve the mining knowledge of underground workers and their own people. Top mining men who became qualified miners through hard work and study should take time and inform their underground officials that it is necessary to carry flame lamps when chemical changes pollute the immediate mine atmosphere. Too many mine bosses have not attended a Mining School, or read a mine textbook and some people think the safety lamps are dangerous tools. I have wondered if the people who are supposed to enforce safety in the pits in the last decade were taught the correct method to examine for gas with the safety light. Too often the flame lights are forgotten for other devices which alarm workers of gas in the immediate atmosphere, but I doubt if coal face machine operators, several hundred feet from the indicators, would hear the warning sounds to let them know the air is flammable. The flame safety lamps will always be the most important tools in coal mines with gas histories. The flame lamp will warn the underground fire-boss before the atmosphere becomes poisonous or explosive, and workers should not be platooned in any gassy mine operations without a supervisor who is using the flame safety lamp.

The flame safety lamp will burn in a gassy afterdamp which is known as "whitedamp"; it is a product of incomplete combustion of smoldering mine fires after blasts. Whitedamp is considered by most mining people as the most dangerous gas found in coal pits after underground fires. It may be possible for experienced mine examiners to suspect that their lamps are burning in an atmosphere which contains percentages of whitedamp, but they may not live to report it in the mine log book. Whitedamp will cause death when breathed for a few minutes. It is mostly carbon monoxide, and it will ignite at the same temperature as methane gas — approximately 1202 Fahrenheit. However, carbon monoxide needs only its equivalent of oxygen to burn inside a flame safety lamp, while firedamp needs ten times more air to burn in the miner's lamp.

It may become necessary after mine explosions to seal one or two sections of a mine deep underground to black-out smoldering fires of coal and other materials on the mine floors. There is a risk during the opening of temporary "banked" mines or sections when sulphureted hydrogen gas makes up uncertain percentages in the closed environments. Small amounts of sulphureted hydrogen may be the meanest of all the gases found in coal mines; it cannot be inhaled without fatal consequences, and it will explode in scanty percentages of air from the heat of smoldering coal in underground sections where fires have occurred. The ignition temperature is only 750 degrees Fahrenheit.

"Blackdamp" is another dangerous gas found in mines after fires and explosions. Although the chief source of this damp is the combustion of carboniferous matter, carbon dioxide may be found near underground water dams. Water which is pumped through small gauge pipes under a pressure from different parts of the mine workings into large mine dams may give off percentages of blackdamp after the water is dumped there and the pressure is released on the volume of water. Although this writer examined water dams many times underground looking for small amounts of CO2 which would dim the flame in my safety lamp, "it never happened." However, I did see the flame lamp black-out from the lack of oxygen in a "sinking workings".

Blackdamp is mostly carbon dioxide, and when breathed it fills the lungs with noxious air and lessens the amount of oxygen needed to sustain life. Carbon dioxide narrows the explosive range of methane gas at explosive percentages, and it is

poisonous in the atmosphere at ten percent, which will put out the flame in a safety lamp. Carbon dioxide is heavier than air and most of the other gases, and over 200 years ago a Belgian chemist distinguished "gas from air". That man's name was Van Helmont. Helmont was a physicist and a writer; he was the first person to tell us that carbon dioxide is a product of fermentation and combustion. The ratio of weight, specific gravity, of air is 1.000, and carbon dioxide is 1.5291.

Fifty years after the flame-closed safety lamps were invented the colliery rescue people were going into fired coal mines and noxious gases "barefaced" to search for comrades trapped underground. The day after England's worst mine disaster in the last century, 27 rescue workers died in other blasts from breathing foul air. The Oak's mine exploded on the 12th of December, 1866, and 334 people were lost in the initial explosion and 27 the next day. The following lines were taken from Britain's mine explosions government statistics. In the year 1891, there were only 51 deaths caused by 23 separate mine explosions, and in 1888 the same number of deaths occurred from 15 explosions. Those years were considered the most favorable for statistics since accurate files were kept on the number of people lost in mine blasts. The largest number of deaths from mine explosions in England occurred during 1866 when 73 mine disasters killed 654 colliers. The average loss of life from mine blasts from 1851 until 1891, per annum, was 223, and during the same period 51 explosions was the average for English mines. Later the writer read the next line or sentence from a skeptical mining engineer on coal dust. "Lately, the impression is, 'rightly or wrongly', gaining ground that many of the explosions mentioned here were the result of firedamp mixed with coal dust." The daily fatalities at the English collieries were four times more, per annum, than those from mine explosions from 1851 to 1891. Mining histories do not list the number of rescuers lost fighting fires, and there may be some mix-up in the dates which the rescue squads wore breathing masks, as it is written that miners were rescued from the Killingworth mine by masked people before the Oak's blast.

British mining history claims that the first closed breathing instruments were invented by an Englishman named Fleuss, and were used for the first time in the Seaham Colliery fire on September 8, 1880. The same man also designed the "Fluess Rescue Lamps", which illuminated the darkness when surrounded by the gases which blacked-out the flame safety lamps. The Fluess Lamps were advanced lighting at that period, but could not replace the safety lamps for the mine workers. Large and heavy iron vessels of compressed air and oxygen were

The St. John Ambulance Association & The Canadian Red Cross Society

This Certificate in

First Aid to the Injured

is awarded to

RONALD McINTYRE

who has attended a course of instruction held under the auspices of The St. John Ambulance Association, at Glace Bay *and having been examined is found proficient in the said subject*

Stanley Scott — LECTURER. INSTRUCTOR.

F. O'Neil M.D. — EXAMINER.

Issued by authority of the St. John – Red Cross Joint Board

DATE: June 6, 1944

Certificate No. 54488

REGISTERED AT OTTAWA, CANADA AND AT ST. JOHN'S GATE, CLERKENWELL, LONDON, E.C.

Ronald McIntyre

carried by one or two of the rescue squad to replenish the lamps and the breathing apparatus. The burned combustion was forced from inside the lamp cases by the air pressure, through a tiny safety valve on top of the light's bonnet.

The supply of air in the first breathing machines for many years was controlled by the people using them; this was not "fool-proof" when something happened to the individual's physical activities in poisonous gases. The modern-day mine rescuer on entering a burning coal pit has about four hours of air at his disposal, which is contained in a receptacle on his back, and his breathing needs are supplied automatically. The gassy damps do not affect the illumination of the electric lamps, and they have electric gas testers, but they also have the flame safety lamps which are lighted after the noxious gases have lessened in the atmosphere.

It may be necessary, when trying to save people and mines, to have miners underground fighting fires, and establish airways that are not wearing closed breathing machines. Safety instructions are given to group leaders and his people before entering a fired mine. Although it is possible that conditions underground may cause a leader to change plans which were made on the surface, he should let his group know about any changes that have to be made. The helmeted rescue squad are advanced some distance ahead of the repair men, and they will alert the people back of them when the gases become unfit, or near poisonous to breathe.

The mine rescue corps is not rushed into an underground fire, and time must be taken to test breathing machines in "smoke houses" on the surface, plot the locations of the fires and plan organization methods. Also, one rescue team should remain near the pit head and be ready to enter the mine if the advanced squads have serious trouble in the gassy afterdamps. Under no condition should a mine rescue person take a closed breathing apparatus from a mine locker-room on the surface and wear it in a fired coal pit without having it tested on his body. Some years ago two noted Cape Breton rescue men were lost wearing such breathing machines which were not checked.

There is a danger entering a mine following a large fire, but the trained and conditioned rescue people have lessened risks, and increased safety at underground fires. The Nova Scotia mine rescue men have been known as "Dragermen" for many years. That name has no significant meaning to these people; it may be the name of a manufacturer in years gone by. Most of the mine rescue teams in Canada are supported by corporations, and mine rescue tournaments are held annually in coal towns.

The chance of a second blast is increased during the opening

"SPRINGHILL" 1956

118 TRAPPED IN COAL MINE

The Halifax Chronicle Herald

and closing of mines or sections of the underground workings when fires are smoldering on the floors deep inside the earth's crust. It is possible for reduced air or have it increased to create an explosive atmosphere when people are opening and closing fired workings. In fighting open fires underground, air samples are taken of the air passing over the burning materials and before the tests show explosive percentages the workers are removed from the fire, and then air currents are closed off in the burning section.

Some time ago this writer had an article published which was headlined "Dragermen", the name by which the mine rescue squads are known in Eastern Canada. That short story described the perils of mine fires and the heroic acts of miners fighting fires in gassy underground workings. There are, or there should be, several passageways in and out of mine sections, but this may be a hindrance to fire fighters, and it is generally necessary to close off all these openings to black-out a stubborn fire. Firewalls are large blocks of coal which are left in mines to contain fires in underground subdivisions; I doubt if they would be of any help when a fire had to be contained in one section inside the firewalls. The passageways are closed as close as possible to the place containing the fire. Safety crews are trained to do this work at fires, nevertheless, there is a danger in the methods through which the air passages are closed. In writing that old paper on "Dragermen", I stated "the first stopping should be built in the air returning from the fired workings, and the last to be constructed, the intake airway." Those lines were taken from an old mining manual, and that method could probably be used nowadays. However, in later mining textbooks, an English mining engineer has written that the reverse of the older method is a much safer approach for the people who perform the dangerous tasks of sealing mine fires. There is much more to sealing off a mine fire inside of a deep coal pit than this writer tells in this book, and it should be understandable, to all concerned people during such emergencies, that the last word of advice of "how and where" a mine or any underground section is to be closed to save lives, is the unthankful job of top mining officials.

The prevention of a mine fire is the sleepless vigilance of a qualified colliery manager, and it is fact that hot, deep coal mines have worked for years without any serious fires. It is also known and regretted that old pits all over North America contain dead bodies of miners who may be alive today if the mining personnel learned more about safe mining methods in the colliery mining classes. Underground ignitions can be decreased in the coal mines by mine officials who have studied from mine textbooks at home. They are the people who are operating clean coal mines for their workers' safety. However, there are people connected with coal mining who think underground housekeeping is a lot of nonsense which adds to the cost of coal. Is it too much of an effort to ask a worker to clean dust from hot grinding gears? Is it necessary for mining machines to strike the roof and pavement several times a shift, and spread sparks of fire into a dusty atmosphere? How often are the overmans people checked at work and told about mine safety programs? Is it possible that safety programs are only encouraged after accidents? These questions can be answered by mining men who know coal pits and can be worked in a safe manner for corporation profits and security for the miners. It is well to remember that coal companies which have not profited from their mines have no security for their employees.

Glace Bay, "A Colliery Community that have become more than just concerned about their future, "But the question remains, "How is this coal to be carried out?"

AERIAL VIEW OF GLACE BAY

PART FOUR

"The Collier's Community"

"MARCONI"

HISTORIC SITE

The inventor of "Wireless Messages", Marconi, constructed the first "Wireless Station" in North America at Table Head, Glace Bay. It was a transitory transmission, as Marconi was building a permanent station on the outskirts of Glace Bay, in the settlement which is now known as "Marconi Towers".

Monument marking site of first Wireless Station in America at Table Head, Glace Bay.

Part Four

"The Collier's Community"

Twenty-five years ago this writer published in the **Cape Breton Mirror** a short story headlined the "Pluck-me Store" which referred to the coal company retail grocery and dry-goods outlets in Nova Scotia, before and after the end of the first World War. This article will cover that story with various accounts not written into the "Mirror" at that time. Coal company stores were a part of mining communities since the first days of mining in England. The United States and Canada have used the same system of credit in their colliery towns, and the company stores became known in the mining districts of North America as the "pluck-me".

The early mine owners and operators preferred married settlers in their mining communities, but the majority of the first miners were bachelors who moved from one place to another, generally searching for a safe mine to work in. Although it may be said that many of the mines were not worked in a safe manner, there were mines and mine managers who enforced safe mining methods. People who were transported into unpopulated regions, to dig coal before mining laws were written, were used up like mine materials which were cast down mine shafts. The first North American miners were rugged people capable of working and remaining underground many hours each day, and they rationed food from "lunch buckets and piece cans" to sustain them working from 12 to 14 hours a shift for their livelihoods.

"Changing times" all over the world have also changed the colliers' thinking, and miners are telling their children, "don't look for work in the coal mines." The coal corporations are aware that a new breed of people inhabit the mining districts, but they are not complaining about the large migration of youths from the coal fields, as the new electric mining machines have left a surplus of people in coal towns for corporations to play with. Nonetheless, the absence of high school boys in the pits will hurt the company and the union at a future date; maybe it has already done that. Mining needs the colliers' "forte" that is only acquired in coal towns.

People find many reasons to leave the mining districts, and miners have found good positions in other industries all over North America, but it is doubtful if the colliers could now find jobs in the steel or auto industries.

Corporations and unions are looking after their own boys

and girls when they leave school, but school counsellors have stated that there are more drop-outs from high schools in industrial towns than in rural districts. Practical mining schools will have to be established in mining towns for the miners' safety in the pits, and their advancement in coal corporation organizations. The schools will help to hold the high school graduates, who are looking for work, in their own towns. However, it would be senseless to think that working people not employed should remain where they were born because an industry may need them in some future time. The provincial government in Nova Scotia has sponsored practical mining schools, and the colliers can attend mining classes during Winter months.

Most of the men now retiring from the mines in Cape Breton had to leave school during the "big depression" years to help support their families. Many were scholarly students who left school with an abundance of intelligent minds that developed in the colliery mining classes. Mine officials now leaving the mines have told this writer that "work and discipline" underground have not lost their respectability, although it might have worked on a more bygone respect.

During the start of the 1970s, it was stated that a great majority of our youths were frustrated with economic conditions, and were delaying the start of their life work, and moving around the country. Many of them did not want temporary jobs, and unemployment became a way of life for people who were trained for white collar positions, and others who were never trained for any kind of work. Although industrial schools have elevated adults in industries, they may not have all the answers for young people looking for suitable jobs in a recessed economy.

Men and women, especially young students who have taken industrial jobs in small or large corporations, must learn that "work and study" are a unit through which people are promoted and their salaries increased. Some of the new mine bosses and factory officials in surface plants think many of our youths have lost incentives for promotions, and they would be better off without that kind of people. However, industrial executives too often place students in positions in large corporations without regard to the youths' abilities, and they have no place to go but down the ladder. People who have started on top positions in companies without industrial know-how will meet the practical students on the "way up" when they start on the "road down". In February 1978, a Canadian college professor told newspaper reporters that some of our colleges should be closed and turned into practical schools and teach trades to people who are now looking for jobs. It is a fact that Canadian manufacturers are importing skilled labor from Europe, and educated people in Canada are unemployed.

Longer than the colliers care to remember, mining towns were off the beaten paths for many people who acquired a little education, as mining and miners were considered the bottom of industrial crafts. Even though coal miners in the United States were once the highest paid industrial workers. Factors in the accomplishment of industrial people have kept workers on the job day and night, and often left mining citizens with no desire to have a social evening with their families or join social organizations. The early mining towns were workshops for coal company employees who had their minds on the "pit whistles", and wondered if there was any work tomorrow. In spite of all that, there were a few families who put their children through school and into college.

Lack of social interests has dashed the hopes of persons in all walks of life, and social scientists have stated that environments and inherited family training was a greater means of success for children in later life than today's school rooms. The **New York Times** criticized that statement by a group of Harvard social researchers. The "Times" stated that some of our greatest people were born in "Hell's Kitchen" on the city's East Side; one of those "kids" became Cardinal Spellman, but they were started in schools with learned teachers. Nonetheless, some parents have given good teachers headaches when their children were not advanced or were chastised in the classroom. However, several colleges were blamed for harboring professors with radical beliefs in the last years of the 1960s, that changed the customary rules of young men and women on the college campuses.

The foundation of successful communities was a combination of good grade-school teaching and the quality training at home; both will show up in adult life. The best people will have nonchalant periods in life and become stalemated citizens, and may have to be motivated by others into community action. Most of the towns and cities in North America were begun by ordinary persons, not corporations, that came to mine the land and build factories, but communities have become wholly dependent on governments and corporations. There are gifted people in every town, and through a little encouragement would start a small business. How many people know that when the common citizen has created jobs for two or more people in his community that this also creates jobs in another town or village?

Just like other towns, mining communities have social

problems which are often aggravated into dissensions by writers who never saw a mine shaft before they were sent to cover a mine disaster or a coal strike which became national news. Most writers think the past is history and should not be hidden from the public. "Tell it as it was" is the motto, but a few will stretch their lines to fill in a column. I've read the following lines from a Canadian magazine (March 1978), "Glace Bay, a mining town, a bleak town with a low, dark sky." The sky is just as high in Glace Bay as it is in any other town in Canada, and I'm wondering if that writer has ever flown in or out of a Toronto airport. However, deep-rooted beliefs, of past performances of others, and abstract problems, still torment mining settlements in North America, and councillors advising residents will have to speak with positive assurance to make popular changes, or they will be looked upon as meddlesome people. Some years ago I talked with a man who told me he was a former Illinois miner who had left the mines after a daily harrassment from others when it became known that his grandfather was a noted company policeman in an old coal company town. Later I read the following story:

"In a Pennsylvania court room a famous United States lawyer, Clarence Darrow, told the Court that coal company policemen on a stormy Winter night drove pregnant women and small children from a company town in the State of Illinois. Two of the women had miscarriages along the snowy road before they reached the next village." It is shameful history of mining which tells that women with chains around their bodies have pulled coal from the faces in the low seams in the Welsh mines, and a boy, not nine years old, was killed in a Nova Scotia mine while about his duties on an underground haulage road. Condonation for mine operators that allowed such wretched happenings to continue for many years is hard to find in mining communities. It is no wonder that colliers are suspicious of outsiders and themselves when they are told "take that," "its' a good thing."

The coal districts have lost the togetherness that John L. Lewis drilled into the local union officers in the United Mine Workers organizations, and the head office in Washington, D.C., is having difficulty with the large districts in West Virginia. During the month of September, 1977, West Virginia miners had completed a seven-week "wildcat" strike over welfare payments which were previously supported from the coal tonnage taken from their mines. In the tenth week of the same strike, solidarity in the union was threatened as U.M.W. members worked while others picketed in District 17. Several thousand Eastern Kentucky union miners in District 30, who had followed the West Virginians on the "wildcat", returned to the pits, but not before learning that the Ohio miners were working. Correspondents have broadcast "the wildcat is over because the miners are starved out." History of colliery strikes have told that miners will lower their standard of living during work stoppages to gain benefits for their livelihood and safety at work. It will take more than a few hungry days to starve miners back underground on a legal strike, but the colliers must obey laws as well as others. Even from the colliers' point of view, this illegal strike was a mistake, and it is possible that union politics allowed the strike to develop in the union halls beforehand, which often causes future events that miners are influenced by at local union meetings.

The colliers are famished from the lack of information from the people who are aware of their problems, but are always late trying to do something about them. It is possible the miners were bluntly told they had to pay $500 of their medical expenses, and without proper instruction at the union halls on why this was happening, the West Virginian miners left the pits under the guidance of their district vice-president Roberts. The union national president Arnold Miller criticized Roberts for the role he played in the "wildcat" strike. Then the vice-president accused Mr. Miller of "soft bargaining" with coal operators, nonetheless, both men were friends and shared time on the same picket lines.

People in Chelyn, West Virginia, are asking each other why they must strike or stop working to have contracts signed and safety enforced in the mines, and benefits, which were gained over the years, taken away without explanation. How often were top union relation people in the local union halls explaining that the cost of medical care had surpassed coal tonnage costs, which lost approximately five tons per man-days worked in the American deep mines in the last two years? The medical deal could have been softened in a "package" arrangement at the end of the present contract around the end of 1977. It is now forecast that the company and union will close the mines at the start of 1978, because of several grievances including the medical deal. It may be a long legal strike in the mining communities of the United States.

Before the last national election in the U.M.W. that elected Mr. Miller as president, it was published that miners who had worked for over 30 years underground in the West Virginia Hills were denied pensions by former union executives. Why did this happen? Mostly because miners who find work in small non-union collieries do not pay union dues, and they were told that they did not qualify for the miners' pension by people in the national office in Washington. Nonetheless, the American

"FORTY CENTS"

Written by "Danny" Boutilier during the 1925 coal strike in Nova Scotia, when each miner's allowance from the United Mine Workers union was 40 cents a week. Mr. Boutilier worked in the mines for over thirty years, raised six children, and was known as the "Pit Poet" in Springhill.

Now a fellow should be hummin',
When he sees the summer comin',
And the little birds are singin' in the trees;
When a fellow wakes up knowin'
He ain't goin' to see it snowin'
And he won't be chilled by any wintry breeze.
I admit I should be feelin',
Just like jumpin' through the ceilin'.
But instead a tear is runnin' down my cheek.
And I'll tell the world, by golly,
It's not easy to be jolly,
When your income is just 40 cents a week!

When there's fifty things you're needin',
And you have to keep on feedin',
Oh, it surely keeps a fellow in suspense;
What an awful strain you're under,
When a person starts to wonder,
How you'll manage when you get your 40 cents.
When you find the rear-end tearin'
From the trousers you are wearin',
And you fear your lady friends may take a peek,
You're in bed while Maw attaches
To your trousers several patches—
When your income is just 40 cents a week.

Almost anyone is able
When they sit down at a table,
To devour 40 cents' worth at a lick;
It is surely some donation,
For to keep away starvation—
Just to think about it makes a fellow sick,
So they cannot call the strikers
In this town a bunch of pikers;
They are men who never show a yellow streak.
It's a cinch that you're not pikin'
If you smile and keep on strikin',
When your income is just 40 cents a week.

miners' pension fund is taken from the coal tonnage produced by coal miners, and many of these coal diggers were paying dues before they were "laid off" at the union mines. There are hundreds of small coal pits in the United States which employ three or four people, but many are closed each year by the state and federal mine inspectors for the workers' safety. The worst mine disaster in North America happened in Monongah, W. Va., during 1907, when 361 miners died in an explosion of methane gas and coal dust.

It may now be past history, but the inner society at the union halls has generally controlled the people who cried out in loud voices for their own benefits, and "fear and favor" was a way of life for some in the mining districts. Emigrant miners were trapped in this for years in the North American coal pits, however, many remained in the coal fields and their children changed the rules. Many years later the writer questioned foreign-born miners that worked in the Pennsylvania and Ohio mines before moving into Detroit, and found jobs at Ford's Highland Park Plant, and it was the same story that my father related to me of the colliers' working conditions during the last century.

The colliers in some districts bought jobs and places in the mines, and even union officers were given choice places underground that were the lawful claim of "run-o-mine" union members. Contract coal face workers had their production and wages curtailed when they lost favor with company officials, and "old-timers" have told that they were days underground for mere pittance, and had to leave and look for work elsewhere. Skilled miners were blacklisted by the union and the company, and moved their families from one district to another looking for work, and were compelled to take jobs in non-union coal pits with fatality rates well above the national average. Not all the foreign-born colliers were handicapped in the mines at Illinois collieries; skilled Welsh machine operators were "big wage earners". These same people would have several miners working for them loading coal at the faces. During the periods that I am now writing about, one of the bitterest coal strikes ever in the American mining districts was started in the State of Illinois.

The district officers rode the "gravy trains" during the years gone by and seats inside were "preferred stock"; the mine committees had fringe benefits which were passed out by the union and the company. National conventions were holidays for a few hard-working miners who were lucky to be included in the number of people sent from each local union. There may have been a few dollars passed around at the conventions as some of the delegates paid hard-pressed grocery bills after they arrived back home; these bills were accumulated from the village merchants. Not all the miners purchased goods at the company store.

The smart mine steward helped others and himself. Notwithstanding, he had the highest paid datal rate paid underground workers or a choice of the best contract places in the mine, lived in one of the best company houses, maybe a few other favors from the company; the "committee man" was the person who was fighting for the workers' equity in the pits. However, many of the oldtime miners reported their difficulties to the mine manager after they were ignored by the underground bosses. The average miner does not want to be known as a person that tells the top officials in the union or the company about certain happenings underground, and it is known for years that miners become "close-mouthed" at serious investigations at the collieries.

However, there have been whispers in mining communities that certain "contemptible bastards" were giving mine superintendents information from union meetings for favors in the pits. During the many hours that I spent with union and company officials, I was never told confidential union news or worked with people who were "tattlers". Probably more so nowadays, people look for extra time or favors in mines and surface plants on their own initiative, and not backward in letting the union steward or his boss know what he thinks.

Some years ago, the union complained to the American Bureau of Mines that gas companies installed temporary underground gas reservoirs in the coal fields, and stated, "gas from these tanks leaked into a mine and was fired." The gas officials told union officers, "if the mines were worked in a safe manner the gas would not have been fired." However, mine inspectors from the "Institute" have reduced this hazard in the coal districts. At the same time it is not always neglect when firedamp burns in mines, although with careful supervision most underground fires can be held at the lowest minimum. There were, and there are nowadays, mine managers who keep their pits as safe as humanly possible for their workers. The miners have praised and appreciated colliery superintendents for their square dealing in placing men in the mines and their families in the company houses. They are the kind of mining officials who never had to be told by government people to "clean up" underground or the mine will be closed. In spite of all that, safety problems have confused the people who are supposed to enforce mining laws in the coal pits, mostly because industrialists have used lobbyists at government offices to relax laws in the American mines.

This story was published, by a top Washington, D.C. news correspondent, during the Spring of 1970. Fifty-two American coal miners have died in our mines because the Bureau of Mines, Federal Government mine inspectors , were not allowed to enforce "good housekeeping" underground. Mine owners took out a court order which would not allow the full enforcement of safety in the mine workings. Corporations should know by now that "housekeeping" in the pits inproves the miners' working conditions and increases coal production.

Miners will lose respect for underground supervisors who allow unsafe practices to continue in the mines until serious accidents happen. Some years back, in a Pennsylvania mine, the afternoon underground foreman was warned about a dangerous roof condition on a haulage road, that he ignored. Later he was told the locomotive operator had refused to move the engine until the ceiling was repaired. The supervisor took over the locomotive himself, and died when the roof collapsed on top of him as he moved the engine through the unsafe ceiling. There is only one cause for any person to do this to himself — "job pressure" — and it is happening every day.

American mine inspectors have claimed that 40 deaths in one year in Pennsylvania mines would have been reduced if the deceased miners had secured their working places before starting to work. Enacted laws written into books by legislators, state "miners must secure their working places before and after starting to mine coal." Too often the penalty for not working in a safe place or manner is death for the collier, but this would not occur if the underground fire-boss had taken time to instruct the worker or remove him from the unsafe working place.

Mine accidents will be reduced and operators' costs lowered with people trained in the colliery mining schools by teachers who know from experience what mine work is all about. There is no doubt that untrained underground bosses are making life in the pits unhappy for everyone concerned with mining operation nowadays, and it is about time that coal operators realize it is going to be that way until mining classes are set up in the colliery districts. Surely it is not the want of sufficient capital in the American mines which is keeping the mine classes out of coal towns.

The United States and Canada are going to become more than just concerned with coal production and the relations between the workers and the mine operators. The dependence on others for energy has become an economic failure in North America at the rate of consumption in both countries above the Mexico border. The Secretary of Energy in the U.S. has told television audiences that something must be done in the coal fields to improve relations in the mining districts with the people who are going into pits to mine coal. The secretary stated that coal must come to the surface "right now" to save oil for future emergencies in the country. In spite of the many work days lost in mining districts during 1977 by stoppages over hospital payments, 128 miners died in the American mines. Canada will use approximately 20 million tons of coal mined in the United States each year.

Concealed information by the union and the company have caused work stoppages and prolonged strikes in the mining districts. Fictitious reporting or rigged tales have left a bitterness in some coal towns which took years to forget it ever happened, and have cost good people their jobs as officials in the union and the coal companies. Miners have listened to extreme radicals and read undemocratic labor literature, and a few have been become known as "kickers" at the collieries, and at times have become officers in the local union. How do the company and the union deal with these people? That may depend on the individual's political power among the miners; at first flagrant propaganda might quiet them at the union meeting, then the company might think he was a good man to have in their organization. Generally the persistent kicker could be suspended for minor infractions; others were transferred to another colliery or asked to take the early retirement plan that was used to remove a local union officer from the company payroll in District 26 a few years ago, and everybody was happy, including "Jake".

The relations, which close diversities among people where seeds of suspicion were sown years before the present inhabitants were born, may be found in halls where men and women gather to have meaningful purposes in their organizations, and that is what a good society is all about. Nonetheless, societies breed indifferences, and bring out the worst and the least of the best people with double-dealing propaganda lectures. People who follow respectable organizations for their own benefits, and using others to help themselves, will not create the harmony that people are looking for in the mining towns.

Social educators, who had roots in Cape Breton Island, have helped the Nova Scotian miners to create new images for underground workers. The colliery people were organized into groups by the late Rev. Doctor Tompkins, a mining village parish priest. The miners constructed their own homes, and landscaped a small community (Tompkinsvilles), near the old "5 and 10" coal mines at Reserve, Cape Breton. Sometime after or

Cape Breton Post—T47

The Gazette
GLACE BAY — NOVA SCOTIA, SATURDAY, OCTOBER 2, 1943.

EXTRA

FOUR KILLED IN SHAFT

FALL 500 FEET TO BOTTOM OF COAL SHAFT FROM CAGE

Knocked Off Staging Over Cage By Heaving Jar But Four Others Are Able To Hold On

GLACE BAY, Oct. 2.—Four men were plummeted 500 feet to instant death in the coal shaft of No. 2 colliery this evening. Four others narrowly escaped a similar fate when the coal tank on which they were coming from the bottom was jarred

WORKING EXTRA SHIFT

GLACE BAY, Oct. 2.— Barely restraining her tears, Mrs. James Stevenson, widow of one of the victims of the No. 2 tragedy, told a Gazette reporter: "If he [illegible] really dead, why hasn't the [illegible] company told me about it? W[illegible] haven't I been told that he w[illegible] in the accident?"

Seated on a chesterfield with Mrs. John Alex MacDonald, wife of the international U. M. W. board member, and a neighbor, Mrs. Reuben Ross, Mrs. Stevenson still clung to the faint hope that her husband might have escaped alive.

"The first I heard about it was when a boy came running over and told me across the fence, that there had been an accident," she said. "Then a neighbor came in, and said he had been in the cage."

Overcome with emotion, Mrs. Stevenson was unable to talk and a sister, Tena Burns, supplied the reporter with the nec[illegible] [illegible]formation. Mr. Steven[illegible] [illegible] had gone out [illegible] her-

immediately a[illegible] pened, but [illegible] sister [illegible]

t[illegible] 12 [illegible] ly [illegible] rive[illegible] said [illegible] four [illegible] comin[illegible] 48 [illegible] enson [illegible] and fou[illegible] Donald, Mary 4.

Also su[illegible] ers Billy [illegible] water, N.S. [illegible] deen, and two sisters [illegible] Tower Roa[illegible] [illegible], Dor[illegible]

CARL ADAMSON WAS HOME FROM ARMY ONE WEEK

GLACE BAY, Oct. 2.— The parents and family of Carl Adamson, 26-year old victim of the No. 2 disaster today, were away in Catalone when the news of the disaster first came and could not be contacted for several hours afterwards. They had left for Catalone at three o'clock this afternoon.

Adamson, who had returned [illegible] Bay on [illegible] home a week [illegible]

ALBERT CHIPMAN WAS SHAFTMAN ON DEATH RIDE

GLACE BAY, Oct. 2—Albert Chipman, one of the victims of the cage accident in No. 2 colliery this evening had been employed as shaftman in that mine for a number of years.

He was sixty years old and a resident of Lower Main street.

Surviving are his wife and four daughters, Doris, Rita, Myrtle and Elsie. A sister resides in Syd[illegible]

[illegible] members of the family were [illegible] are of the tragedy that [illegible] their home until they were [illegible] d by The Gazette [illegible] fr. Chipman [illegible]

Tom Foley Had Worked in No. 2 For Past 43 Years

GLACE BAY, Oct. 2.—"There's nothing too good that can be said about Tom Foley," a neighbor told The Gazette reporter as he entered the Foley residence in New Aberdeen. "You could ask questions about him for days and all you'd hear would be good."

In the two-storey residence, Mrs. Foley was grouped among consoling friends and relatives [illegible] had arrived. Her condition [illegible] was unable to

said Patricia. "Dad was late coming home."

Douglas, the youngest boy in the family, sat on the bed with his brother-in-law. "Dad worked in No. 2 ever since he came to Canada 43 years ago," he said. "He's still got his first check dated 1900."

The account of Mr. Foley and his family was told quietly, and there were no tears. But every once in a while, the conversation would die down, and the reporter could sense the anguish felt. [illegible] Tom Foley [illegible] years old. [illegible] New-

Murray Graham Gra[illegible] Coat Tail and Held O[illegible]

GLACE BAY, Oct. 2.—"It happened so quic[illegible] know what happened", Murray Graham, one o[illegible] escaped on the death ride at No. 2 colliery [illegible] just stated.

"I was standing on the tank [illegible] [illegible] we were about half [illegible] during the period in which the miners were building their own houses, the Rev. Doctor Coady founded the "Coady Credit Union"; an organized number of common workers were going to loan money to each other without bonded collateral, at low interest rates. Both of these clerics were born on the banks of the Margaree River, a salmon fishing stream where this writer spent many happy hours, but the two "Fathers" were always too busy helping others to make or cast salmon flies.

Mine accidents are generally reduced during slow production periods at the collieries, and John L. Lewis told government representatives at one time that the Summer strike would prolong the lives of 300 American colliers. That was the late union president's concept of mine safety in the pits during the war years; now we are told more miners are maimed in the mines than ever before. Nonetheless, the United States federal and state mine inspectors, with some participation from the union and a few coal companies, have greatly reduced the yearly fatality records in their mines.

The enormous amount of coal taken from the crust of the earth during the last half of the 1960s has contributed to a profound loss of people from the mining communities around the world. It may surprise mining historians that 1965 was the most unfortunate year in coal mining, in regards to the loss of miners in the pits, for many decades. The actual number of colliers who died in the mines during the sixties all over the world will never be counted, but in the space of less than three weeks in 1965, more than 750 underground workers died in coal mines.

Most of the large coal corporations in North America have reduced deaths in the coal pits with strict enforcement of mining laws, although too often workers think they are subjected to harassments when checked by a mine official. The total number of deaths reported from the American collieries for two years, 1975 and '76, was 295, and this must be considered a merited reduction for mine safety in the United States. During 1967, that year's fatality count was 217, and that was the lowest annual death-count ever in the American coal industry until the years mentioned in this paragraph.

Mine laws should be administered with authority underground, that may be impossible for unlearned mine bosses to execute after they lose the respect of the workers. Teaching people to work in a safe manner was always a greater means to accomplish the workers' confidence in the pits because they realize the boss is interested in their safety at work. The supervisor who instructs others in a firm and friendly manner is the kind of boss needed in the mines. The old mine boss's motto

The fundamental importance of hospitalization had become more apparent at Glace Bay with the increase of new mines, and concerned citizens constructed the second hospital in Nova Scotia during 1901. The first hospital in Cape Breton was at Louisburg during the 1700s.

ST. JOSEPH'S HOSPITAL – ERECTED 1901

GLACE BAY GENERAL HOSPITAL – OPENED 1914

"hit them in the pocket" must be forgotten, nevertheless, the pressure that keeps people working safely should not be relaxed.

The two greatest mine disasters were 36 years apart; the first in France during 1906, when 1000 colliers perished underground after the "Courriere Colliery" exploded. Since that catastrophe, France has the lowest fatality rate in the world of mining coal safely, however, they were never large producers like England and the United States. The most destructive coal mine explosion ever occurred in Manchuria, during 1942, when the Honkeiko Colliery exploded, and 1,549 miners died in the blasts. It has often been asked why so many people were underground at one time. It was because of the dangerous practice of changing shifts underground, and if that is happening today, it should be stopped. Some years ago at an American mine inspectors' convention, it was stated that during 1907 more than 1,000 miners died in three colliery explosions in North America. The writer could not find a complete report of the disasters mentioned in the last sentence, but it is a fact that in the month of December 1907, two mine explosions killed 588 American miners. England was losing 1,000 mine workers annually for decades, and it was not until government mining schools were placed in the colliery districts that fatalities were notably reduced.

Nova Scotia is a small coal-producing province in Canada, yet there were many tearful scenes near the pit-heads after underground explosions since 1873. Most of the fatalities which occur in mines are singular accidents that happen too often, and there is not much public knowledge to explain why or how they occur with such frequency. Most governments do hold enquiries, after fatal accidents at the collieries, attended by the union personnel and the company mine officials. The State of Virginia is also a small coal-producing area in the United States, but in the first two weeks of April 1978, nine people had already died in their mines.

Gas and dust explosions in the Nova Scotia mines were reduced since the first World War, through the introduction of the colliery mining schools which the provincial government took over and supported. There is no doubt that most of the mine blasts in Eastern Canada during the early years of mining were the result of inexperience in gassy coal seams. Sure, it has been stated that many of the older miners did not fear small amounts of gas that collected at "room-ends". The early miners played around gas pockets with open flames of fire that often cost the lives of themselves and others.

Most of the people who arrived in Pictou County and Cape Breton from Scotland in the late 1700s and the early 1800s were

NEW ABERDEEN ---------------------- GLACE BAY

Coal Mines

NO. 2 COLLIERY

NO. 26 COLLIERY

those who tilled the soil and tended sheep in the Highlands. These people never intended to work underground for their livelihoods. Their descendants became skilled miners through experience and the colliery mining schools, and one of the first mining school teachers in Nova Scotia was a man named MacKenzie from Pictou.

The first mine disaster to be recorded in Canada happened in Pictou, Nova Scotia, during the month of May, 1873, when the Drummond mine exploded and 60 people died in the fire. The Albion Colliery in the same district had a history of fired gas before it had a major disaster in 1880, that killed 49 miners. The courageous little mining town of Springhill, N.S., had its first major underground affliction, during 1891, that took the lives of 125 colliers. The first mine explosion on the Cape Breton Island was at Sydney Mines during 1878, and six coal miners were lost. The Vale Colliery on the mainland had a blast of fiery firedamp in the month of May, 1881, and seven miners were killed; four years later the Vale mine had another explosion and 13 people died underground. During the middle of the last century explosives were stored in the mines, and from 1866 until 1872 there were 40 miners killed in the pits in Nova Scotia after powder burned and exploded underground. The Caledonia Colliery at Glace Bay, N.S., had an explosion of methane gas during 1899, and 11 men died fighting the fire. One of the victims was the underground manager of the Caledonia mine at that time. One of the collieries at Reserve Mines in Cape Breton during 1903 exploded and there were five fatalities. The Port Hood mine in the County of Inverness, lost 10 miners when firedamp burned and exploded in 1905, and that was the last recorded, major mine explosion in Cape Breton mines until July 25, 1917. The Number 12 colliery exploded that day and 65 New Waterford residents died in Cape Breton's worst colliery disaster.

The miners who worked in the "Allan Shaft", Stellarton, N.S., in years gone by, have told stories of gas and water which stopped the coal production from one of the world's richest coal seams. The "shaft" coal is over 40 feet thick; it might be the highest in the world, although a mine in Poland has claimed that distinction. During January 1918, the Allan shaft mine exploded and 88 died in the blast. Before and for a few years after the first World War mining in the "shaft" was crude and gassy. The miners worked on ladders in the high coal to scale the roof, or to pull down coal that hung to the face after the shots were fired, and it was told that the men on the ladders were often seriously gassed and fell to the mine floor. The MacGregor mine in Pictou County exploded in 1952, and 19 were killed, and in Cape Breton, at New Aberdeen, when a small body of firedamp was exploded in Number 20 colliery on the night of July 9, 1952, seven colliers died in the blast. It was suspected that the explosive methane was ignited by a leaky safety lamp.

The town of Springhill had a major mine disaster on November 1, 1956, and 39 miners died in an explosion of firedamp inside Number 4 mine workings. Two years after that unfortunate happening, Number 2 mine at Springhill which was subjected to minor "bumps" for several years had a series of several quakes underground that killed 76 miners on October 23, 1958. Both of these tragedies have shut down deep mining in Springhill, which started in the middle of the last century. The coal seams beneath the town of Springhill have approximately 4,000 feet of cover, one of the deepest in the world, and surely the most difficult to mine. The "Springhill Story" has filled news columns, and a few books, and most of it is mining tragedy.

The late Thomas J. Casey presented a paper on "bumps" before the Mine Inspectors' Institute of America, when he was Chief Inspector of Mines for the Province of Nova Scotia. Mr. Casey stated " 'bumps' is a manner of speach used by miners when a sudden breaking of the mine strata and upheavels which crumps stone and coal underground, moves heavy machinery, and often destroys the mine airways and the miners' passageway to the surface." Casey continued, "The Springhill mine bumps have had a thorough study over the years, and their is no doubt in the minds of geologists and mining people who have examined the data concerning the Springhill coal beds that the problem is the thickness and strength of the sandstone strata above and below the Number 2 coal seam."

Most of the bituminous mines have methane gas located in small quantities when room mining is allowed to advance beyond the pressure which moves ventilating air currents into the coal faces. The gas which accumulates in "headings", "off the air" is generally removed safely by experienced miners under the direction of the section boss into the return airway, but that gas may not be reported in the mine log. It is also known in mining towns that both workers and mine officials have concealed information when questioned on gas pockets that were brushed from the coal faces by unnatural means until a "cross-cut" was mined through the pillar to clear the air. Nonetheless, the section deputy should not "put up" with ventilating problems which were caused by careless mining people. Presently, in most large mines, "booster fans" are used in "tight" places to keep extra long headings gas free.

There are erroneous statements given during enquiries of how a fire was started in a dusty mine roadway. The falsity of which may be found in a coal miner's textbook. The day should be long gone when such vital matters that are affecting the safety of miners and the people who work with them are left in the hands of mine bosses who are not capable, through their mining ignorance, and it should not be tolerated by any government. It is also essential for the local union officers to acquire more information upon the explosiveness of coal dust in non-explosive methane gas that miners are breathing in the mine air. The place to get this information is in the colliery mining textbooks, when there are no practical schools for the miners to attend. However, too many colliers think that a short study of mining methods is no concern of theirs, yet they are breathing air that is polluted in the mine workings, and somehow, someone with a little mining knowledge may find help for the miners before they are afflicted with "black lungs after ten or fifteen years working in dusty coal mines."

Sure, the people who control mining operations take some comfort from the reduced fatality count in the mines after explosions in North America, but they are still happening too often. The reduction of people in the pits after mechanized mining machines were introduced is one reason the fatality count is lower in the American coal mines after explosions. That statement is only partly true because the constant surveillance in the mines by mine inspectors has decreased the number of ignitions in the coal pits around North America mining districts. However, daily accidents must be curtailed in the mines, and the only real help for the people who are trying their best to curb fatalities is through the colliery mining classes.

Mine disasters were newsmen's headlines long before the mishaps that travellers now endure in aviation tours, and information on the causes of these unfortunate accidents in the air and underground is just about impossible, although people have plenty of misinformation. The mine owners did not know what had happened underground; they were never down there, but they had a story for the news people from their offices. The top mine management and the engineers who spend most of their working days on the surface had an idea why the mine was fired, but they were not talking, yet they knew that most of the underground bosses never read a colliery mining manual, nor did they ever instruct their firebosses in methods of mining coal in a safe manner. The local colliers also had information for newsmen, but at the investigation of a mine blast, could not explain where or how the gas accumulated and how it was exploded. The apparent changes that people make, when questioned on the causes of accidents, are because the truth may do more harm than good to the individual who would be often ridiculed.

There were and are coal mines which operate without a large number of accidents, and there are mine officials in the pits who operate without gas problems in gassy coal seams. The trained mine manager will inspect the underground mining advancement and the quantity of air needed for each working section in his mine. At the same time the mine manager or superintendent must have reliable information and explanations from his mine supervisors to operate a "clean" coal pit. The miners have, for years, argued with coal company officials over payments, but seldom have they stopped to think about their safety underground in the past decades.

The miner knows who he is; he is one person in a collective working force who will go down in mines for a livelihood, but others must remember he needs a little more than just a good livelihood, and he is willing to work in the crust of the earth for other people's needs as well as his own shelter. The miner never claimed his service to be meritorious, but it was many times.

"Dirty coal mines" are the graveyards of numerous mine workers of the past, and miners' bodies have rotted in them because mine owners were allowed to mine deep inside the earth, and they did not have trained people or the equipment to fight large underground fires and for rescue work after explosions. Miners who became trapped in coal pits after fires and blasts have lived for days waiting for help that never arrived, because nobody knew how to help them, although some have tried and also died in the gassy damps.

There are several miners' museums in the United States which are supported by people in the coal trade, but colliers' practical mining schools are hard to find in the coal states. Some years ago the colliers were supplied with books and correspondence courses, and the "McGraw-Hill" publishing company sold mine textbooks in the United States and Canada.

"Black faces" will not maim healthy people who are working in "clean coal pits"; it is the bad air that miners are breathing in dusty atmospheres that causes health problems in the colliery communities. However, most of the people in the mining districts think it is one of the work hazards that miners have inherited and nothing can be done about it. The miner will have to have clean air to breathe and let's hope it is not too late.

The colliers have played baseball and other sports during the long evenings since the eight-hour day became a shift in the mines. Now, the average collier is in bed at twilight, resting for another shift, breathing air from an atmosphere which contains silica and noxious gases. Silicosis is a sickness which became widely spread in coal towns since the 1940s. It is a known fact that more than 100,000 North American mine workers have contracted black lung since the miners started to work in high velocities of air with new machines powered by electric motors. Have coal operators and union executives ever wondered why old-time miners who had worked underground for 40 years digging coal in gassy pits which were poorly ventilated, and were not seriously injured during that time, did not have lung disease? Some of these miners worked in the pits until they were 70 years old, and lived for another decade without company pensions. John Devoe started to work in the coal mines during the last century, and passed away in February 1979.

The effects of daily accidents, black lung, and discontented mining communities have now become critical and will worsen if strong and determined people in coal towns "don't start pulling on their own boot straps." It is hoped that the colliers have not lost the intent for reforms since "Father" Tompkins' days in the mining communities of the Cape Breton coal fields, and will have to be motivated into action to help themselves

John Devoe

NORTH SYDNEY — John Devoe, 95, of RR 3, George's River, Cape Breton County, died Friday in Northside General Hospital, North Sydney.

Born in the Magdelene Islands, he was a resident of Cape Breton, most of his life, working in all of its coal mines.

During the Fall of 1976, a Canadian cabinet minister was speaking to a group of miners who were reluctant to give him any information, when he stated, "Tell it to me as it is and do not be afraid of reprisals." Reprisals from whom, the company or the union?

The State of Indiana has a Mines Act similar to the writing on the coal mine act of Nova Scotia; both were taken from laws in the British mining industry. However, acts and laws have to be updated from time to time, although seldom is the run-o-mine worker informed of any changes. Most of the countries which mine coal on a large scale have mining laws to protect the workers and the wealth of the coal seams. At the same time union officers should beware of modifying laws instituted to increase production underground.

Coal mine operators are searching for skilled fire-bosses (overman), and the last advertisement that this writer read in a daily newspaper stated — experienced fire-boss needed, starting salary $20,000 a year; I did the same job during World War II for $175 a month. Coal mining has a future for youngsters in the colliery towns; that is where the money and action will be in the future decades. During the years gone by, miners who studied in the Nova Scotia colliery mining classes were often offered jobs in the United States and Western Canada, and some were in South America for several years. The United States should have mining schools in their colliery districts, and I talked to a mine superintendent about this, and he replied, "We have engineers from our colleges," but how many fire-bosses are mining engineers? Down through the years at the collieries, engineers and colliery school graduates in the pits did not

Nº 2920

Province of Nova Scotia

Department of Mines

This is to Certify that RONALD HUGH McINTYRE of GLACE BAY County of CAPE BRETON now the holder of a MINE EXAMINER'S Certificate having duly complied with the provisions of the Coal Mines Regulation Act, is hereby granted a Certificate of Competency as a

OVERMAN

Given under my hand and seal of office, at Halifax, this 5th day of JULY A. D. 1940 always see "eye to eye" on mining problems, and would have bitter exchanges before adjustments were made underground. And that same view is held in most other industries, but in mining, people who became underground supervisors must have proven their mettle on the job. There is, however, still evidence that many underground officials are only "coal pushers" in the pits.

Most people know that coal miners are just like the family next door because mining towns have many visitors during the Summer months. Nonetheless, there were a few ignorant writers who have written, "Miners are unwashed and illiterate," and they are responsible for the loss of their own livelihood through strikes which closed mines. This absurdity would have never been written if the writer had talked with the coal town residents before he published lies about them. Many of the first people who worked in the early coal pits in North America were foreigners with a language barrier. In new communities, most of them were neither illiterate nor dirty. Cleanliness is more than just washing his hands to the collier. The miner bathes every day and prays at night for his safe return to his family after the next shift in the mine. The Polish miners who immigrated to Cape Breton at the first of the 1900s washed their pit clothes every week, and I know miners who changed their "pit socks" every morning of a working day.

Values were lost and saved in the coal districts during forced economic idleness and counteracting circumstances. Some people were off work for months due to sickness and mine injuries, many who had no help beyond the charity of neighbors in the company houses. There was a mine disaster in Kentucky not so many years ago, and the company claimed they had no money to pay conpensation to the victims' families. Some mines only worked one or two days a week; others did not produce during the Summer months, while still others stopped mining during the Winter months. There were miners in Indiana who only worked in the Winter for years, and at one time, in Kentucky, mine operators were paying miners with "scrips" that only had a value in the company commissary.

Late in the 18th century, a coal mine and a company commissary were operated on Cape Breton Island by two Halifax businessmen. The small settlement became known as the "Mines". Although the two men were partners in the mining venture, one was a merchant who operated the store for a profit and his associate was expected to do the same at the mine. However, the commissary was operated in a manner which would be considered a deterrent to mining supervision nowadays. That old company store sold rum to the miners every working day. The first "pluck-me" was built near the mine mouth, and during the long shifts worked at that period in shallow coal pits, most of the miners came to the surface and ate their lunches in the store. It was a habit for the underground workers to call in at the store several times a day for supplies and their meals, and have one or two drinks of rum. Sunday was the only dry day of the week, and most of the miners found a spot to worship God. Rum was drunk in Nova Scotia during the 1800s like the people in France used wine on their dinner tables. It took an awful long time to break this habit of drinking every day, but now children see their parents consuming liquor on Sunday afternoons.

Stout and Tremaine, from Halifax, took a part of their business into Cape Breton for a mining operation. Mr. Tremaine operated the store, although it was his money that paid for the coal lease. Mr. Stout was the mine boss, and he must have complained to his partner about the liquor sales to his workers during working hours. The store manager sold tools, clothes, groceries, mine work boots and rum, which accumulated into a long list to be paid through the payroll each pay day which happened once or twice a year. Then the young miners moved to another mine.

The early colliers at the "Mines" entered the mine at five o'clock in the morning, and their first job was to mine the coal face and make it ready for the day's production. The mining and loading of coal into small boxes was completed by "hand work". In later periods the coal was filled into mine cars which ran over wooden rails. The mine workings were only a few hundred feet from the surface and at nine o'clock most of the underground workers took off for the company store to eat their breakfast there in a group. During breakfast, which took about a half hour, most of the older men had one or two drinks of the "well" watered rum that Tremaine sold at high prices, before they returned to work in the mine. During that time period in Nova Scotia, rum cost less than molasses on the Halifax waterfront, and city clothiers served free liquor at their doors to stimulate friendly customers into buying their wares. Tremaine would "perish" if somebody got something for nothing in his store, and rum was a fast-moving part of the store's merchandise.

At 12 o'clock the miners arrived back in the store for an hour and had their midday meals and drank more of the merchant's watered rum before returning to the shallow mine workings. More than likely, one or two of the colliers missed the last part of the shift on certain days, and were promptly disciplined the next morning. The colliers were given whatever the store had in stock, as Tremaine was the mine manager's bookkeeper, time-keeper and paymaster; he had first count before pay days which were once or twice a year during those first mining years at Sydney Mines.

It was the same "goings-on" in the store at the Mines when the shift was over for the day, seven o'clock at night; after the workers had their stomachs filled they talked about the amount of coal each mined and loaded that day, as all wanted to know who was high-man on the tonnage sheet. Some had books, and others read from letters that arrived in the district once or twice a year. At night these first miners slept in cubicles which were kept hot Summer and Winter. Sunday, during the Winter months, was the only day of the week that miners, who had their lunches close to the coal faces, looked at the sun overhead. During the Summer, on Sundays, the miners rolled up their pants and removed their socks and boots and sunbathed in the warm sunny atmosphere on the surface.

Christian names are often forgotten among the Highland people in Cape Breton, and a name relating to a happening or an act was attached to many families in the years gone by. Most of the people didn't mind, some did, and strangers had to be careful when mentioning names during conversations. However, it was seldom that a pseudonym was blanketed over a community, but a nickname which is long forgotten was "hung" on a mining settlement in Nova Scotia over a hundred years ago, "the sun-dogs", so-called for bathing their feet on sunny Sundays in the colliery district. The first collier's clock was the "whistle" in the pit yard, and the miners worked, slept, washed and cursed to its ear-splitting screams. Early in the morning the whistle announced a working day, and at night "no work tomorrow".

Before the days of washrooms at the Cape Breton collieries, the miners washed up at home immediately following a shift in the mines, generally in the kitchen. Some families had porches built off their houses which were used for washing clothes and bathing. A large wooden tub was filled with hot water near the coal stove, and with homemade soap each member of the family who worked at the mine scrubbed himself; as the tub was refilled with water from pots on the kitchen stove. Then the miners dressed in street clothes and waited for dinner in the late evening. The kitchen was off limits for the women when the men were washing themselves, until "Pa" shouted, "Ma, get the dinner on the table." Then the mother of the family would peep through a curtain door and tell the girls, "Come on, get the place cleaned up and the dishes on the table." The same wooden tub was used on the weekends to soak pit clothes which were hung to dry out of the sight of gossipers — "Mama's washlines had to be snow white, as clothes worn in the sweat and dust underground never returned to their natural color."

Fresh milk, butter, eggs, ham, mutton and other items were not a daily diet for the early Nova Scotia miners, although people from the surrounding countryside sold these foods to the colliers during the Summer. However, it was not uncommon for a family to have a goat to milk during the day. Cape Breton miners have followed for years a menu of oatmeal, potatoes, fresh and salt fish; corned beef and pork were pickled in the same barrel several times a year. Sunday dinners consisted of corned beef and pork, some vegetables, and large "doughboys" with molasses over them, and for dessert, gingerbread and cups of strong tea to be drunk listening to the Caledonia pipers.

Sydney Mines became the first mining cènter in Cape Breton during the 1800s. The coal seams "cropped" out of the ground in the district, and the town was built on the north edge of the Sydney coal basin. My father visited the "Mines" during 1868, to attend the funeral of his cousin who was fatally injured in the mine. Then, as a teenage boy, he took off for the United States where he followed mines and mining towns for 15 years, searching for an older brother in the American Midwest. Changes were made at the mines when he returned to Canada, and the General Mining Association had taken over most of the coal leases in Nova Scotia. In Cape Breton, the Sydney coal basin extended from Sydney Mines to Port Morien, but the mining association did not control all the mines in the Sydney mining districts until the industry was taken over by the Dominion Coal Company. The Caledonia colliery was privately owned until 1893, and the Mitchell family from Halifax had a coal charter in Dominion, but they were connected with the Mining Association, and sold their rights to the new company.

The new company bought mining machines and changed mining methods, placed mechanical ventilators on the top of mine shafts, and were the first to have locomotives underground in North America at Number 2 Colliery, in Glace Bay. The strange part about the coal leases was that another company was given permission before the first World War to open a mine in Broughton on the Harbor seam. Nonetheless, the Dominion Coal Company held the lease for the same bed of coal. The Broughton project was a failure, and the principal stock holder was found dead in an elaborately-designed mining town hotel that he constructed to entertain guests. Broughton became a training center for the Canadian army during the first World War.

The new company built houses and company stores in the mining districts and promoted their employees in mining and their other business transactions. Two Glace Bay residents, D.H. MacDougal and H.C.M. Gordon, held offices at different periods of president and general manager of the Dominion Coal, when it was the largest mining company in Canada. Harold Gordon was a graduate from the Glace Bay High School with honors, and later attended a military school. He became a mining engineer and as a young ambitious collier took over the construction of the motor road at number 1B mine. Later, Gordon was appointed the residential engineer at Sydney Mines; his mining ability and the trust which the coal corporation placed in him were never questioned. However, Mr. Gordon had more than mining knowledge in his composure; his family training in the colliery districts of Cape Breton gave him charity for the unemployed at the old Sydney Mines collieries. During the depression years, one of the mines under the engineer's administration was closed for the Winter; there were no unemployment cheques at that time, and that Christmas, Harold Gordon, then General Manager of Coal Operations, advanced on his own initiation $100 to each miner unemployed from the stricken colliery.

The Princess Colliery, at Sydney Mines, had a terrible underground transit accident on December 6, 1938. Twenty-one miners died when a "riding rake" cable broke during the transportation of workers into the mine workings. That disaster will not be forgotten and the union blamed the mine officials, and said, "Some company bosses knew the rope which broke at the most unfortunate time was 'well worn' and should have been changed before the disaster happened," and the General Manager became a target for Adam Scot, a union district officer.

The disastrous mine explosion which occurred at Springhill, Nova Scotia, during 1956, was softened a great deal by the rescue of 88 miners who were trapped in gassy air for three and one-half days. The rescue was by all human standards a "miracle", but the town of Springhill had a mining phenomena during the "Bump Disaster" in 1958, when 11 colliers were taken out of a mine after nine days underground, by the Nova Scotia mine rescue people. The rescue operations after the 1956 explosion in Number 4 pit were handled by H.C.M. Gordon, then the general manager of the Dominion Coal Company, and Thomas J. Casey, the chief inspector of mines for the Province of Nova Scotia at that time. This writer was told the mine was "banked", temporarily closed to smother fire, before the rescue of the 88 miners several hours later. However, I learned from authentic sources that the slope was not closed until after the men were rescued. Nonetheless, there was a conference in the mine office, and the discussion was the "sealing off" of air going into the burning underground workings, when the chief inspector asked for another look inside the mine slope, and 88 men were found alive and removed to the surface. Late in the Fall of 1909, the Cherry mine, in the State of Illinois, burned and 259 miners died in the fire. The pit shaft was closed under the protests of people who gathered in the colliery yard. After seven days the shaft was opened and 12 men walked out of the burned mine workings. It was made known later that several Nova Scotians worked in the Cherry mine during that terrible disaster. During the "big strike" in 1909, miners left Glace Bay for the coal fields in the States and Western Canada; the Morrisons from "Steele's Hill", for example, moved to Iowa, where they had relatives.

Lack of Winter shipping from the two small harbors at Glace Bay coal districts practically closed the mines during the first days of mining in the Winter season. The ice-bound water off the coast caused long periods of slack work weeks or no work, and most of the town's residents accumulated debts that were paid in the Summer when the mines started to work full-time. The miners had the company store, and the "pluck-me" did give credit to the people employed at the collieries during slack periods at the mines, and the colliers who took advantage of this deferred payment plan had the next Summer to pay the bills collected during the Winter. The village merchants gave some credit to the people not employed by the coal company, and maybe to a few miners that company store clerks thought or were informed to be good "credit risks". The contract mine workers were capable of "paying up" in a couple of months when the pits started in the Summer, but it was a long and hard pull for the colliers who worked on a low daily wage to reduce their "back-bills" in the "pluck-me" store. In spite of all that, there were families who saved a little and always had a few dollars at home, and a few miners educated their children in Nova Scotia colleges.

The mining village of Glace Bay became a boom town at the start of the 1900s, and people from all parts of the world arrived there to work in the mines or to establish a small business. Glace Bay became an incorporated town in 1901, and holds the distinction of being the first town in the British Empire to receive from Edward the VII its official charter which guaranteed its inhabitants protection under the British North American Act as long as there is an England.

Glace Bay at one time was surrounded by a dozen coal mines that employed approximately 8,000 people, and most of them and their families visited the town center on Saturday evening. The streets were crowded with people who conversed

GLACE BAY NATIVE PTE. JOHN BERNARD CROAKE
VICTORIA CROSS AWARD, WAR 1914 - 1918

in several languages for three or four hours before the company train left for the outlying districts between nine and ten o'clock. There were many liquor vendors who sold "booze" under the old "Scott Act", but the old "rum-sellers" were harassed by inspectors who warned them not to be caught "selling", but could give them and others all they could carry for nothing, and it wasn't always liquor. Nonetheless, most of the "bars" were a legal business.

The bars were supplied by liquor salesmen, from Montreal and Halifax, who displayed all the brands found in the government liquor stores today in Nova Scotia. The "beer salesmen" were the "big spenders" in the mining towns for several days before they returned to their home base, and they were the last of their kind in bars for fellows who waited for a "live one" to come in and buy drinks for the house. Three years after the first World War started in Europe, the Nova Scotia Temperance Act became a law in Cape Breton, and "rum-row" was established off the Eastern Atlantic coast. Glace Bay was considered a wide-open mining town as far as liquor sales, "nothing else"; the town never had a house where prostitutes entertained men.

The coal company stores became known as the "pluck-me's" in many mining towns in North America, but the writer was told the name was coined in England. During the last few years of the company stores in Cape Breton, they were called the "grab-alls" in Glace Bay. It was the opinion of some people outside the mining districts that miners had a negative view of their "pluck-me back-bills", and the company should forget them. Not so, but maybe a few thought that company store accounts in their names and numbers could not be paid in their lifetime, and a few have passed out of this world a debtor on the "pluck-me" credit books.

The back-bills were not lined out when some people lost a wage earner if there was a potential worker in the household. That person was given a job somewhere in the coal company's organization at a later date, but that boy or girl might be asked by a store manager to pay his or her father's bill in the "pluck-me". Women never worked in the mines or around the pit-heads in Cape Breton, but girls worked in the company store and the corporation offices. The widow without a wage earner in the coal organization was told, "sooner or later", to look for a house clear of the company rows, as new miners were looking for houses for their families. These women learned about the company policy beforehand and soon moved out, and they did not ask for store credit after their husbands were fatalities in the mines.

R. A. SMITH'S HOTEL DINING ROOM — 1896

PARK HOTEL — 1896

The easy-buying method today with credit cards was similar to the miners' purchasing system during the years the company stores were operated by coal corporations. However, it has to be admitted that the coal company never sent letters to its customers to return and buy more after their accounts were paid in full. Sure, the company store credit system worked for years in the mining towns and villages collecting money from workers' pay envelopes for groceries and dry goods; some blessed it and others cursed the system on payday, but they were not paying compound interest on back-bills because of slow payment.

The company stores supplied their customers with the best brands of food and dry goods, and generally the grocery department was sold out on the weekends. Bread from the Sydney and Glace Bay bakeries arrived every day; that is not happening today in the supermarkets. Payments could be made on the installment plan. When a young man asked for his first suit of "long trousers" at the company store, he would not wear the suit because the older boys kidded him on gaining his puberty. Most of the Glace Bay miners wore tailor-made suits to church on Sunday, as more than a few of the first Scots who arrived in Cape Breton were weavers and tailors. Some of the Jewish immigrants who found a home among the early settlers in Cape Breton were professional tailors from Russia and Poland. Even today, in the towns and villages where Scots from Nova Scotia families are catalogued through nicknames, "the tailor" has followed people that never sewed a button on a shirt.

It would be wrong to say, because of large back accounts not paid in full at the company store during long periods of steady work at the mines, that the colliers were refused credit, but some articles were scratched from the weekly order, although no "junk food" was asked for on the written order. Molasses was substituted for butter, and some were told to have time on the weigher's coal sheet for their next delivery. It might be correct to think it was in the best interest of the coal company to supply credit to customers who worked at the collieries at low daily wages and turned coal production into corporation profits. It has been stated and written that coal companies used the credit system to keep miners in the pits, nonetheless, the stores operated at a profit like any other business.

There were numerous deductions on the miners' "pay-sheets" which were taken from the colliery office a day or two before payday. The "off-take" from the miners' gross earnings increased as the years passed, and several decades back, the colliers were paying through payroll check-offs, taxes, hospitals, doctors, churches, union dues, weighers, house rent,

CALEDONIA STORE

This picture was taken some yeats after the company store (which served the Caledonia colliers) was moved from Brookside Street, near Boutilier's Corner, to Pit Street.

fuel to heat company-owned houses, the current company store account, and there might be a "nagging back-bill" at the pluck-me to pay a few dollars on. The colliers named the pay-sheets "bob-tails". Then after all that, at intervals, a young colliery office clerk read a statement to his chief office cashier, a statement from the "pluck-me", which was higher than the miner's wages, and the company stores in Glace Bay had another name, "the grab-alls".

It is well to remember when telling these stories that there were several thousand miners in Cape Breton, during the company store years in the mining districts, and many of these people paid cash on weekly installments for furniture purchased in the company stores and other merchandise not connected with the coal company. The company store sold goods to non-employees for cash, but would refuse items to outsiders when stock was low in their warehouses. It is doubtful if the company stores gave credit to people who were not employed at the collieries.

My father was refused a barrel of flour and a sack of oatmeal, with the money to pay, at the Caledonia company store many years ago, mostly because he operated a business in the district. Then he had to go into Sydney with a horse and sleigh through banks of snow for the articles he needed. The provisions included a 30-gallon keg of black Demerara rum. Although the other kids called us "rum-sellers", my father never sold liquor illegally, and to buy a drink at his bar a young man proved his age with a good-size moustache.

"Papa" bought our farm from the General Mining Association, during the late 1880s, before any mines were in the Number 4 district of Glace Bay. The name Mitchel was on that old deed; he may have been one of the men who opened a mine on the shore near Table Head or the Roost, as this part of the town was known at one time. The Caledonia mine officials undermined our land at a later date, and also took a strip of our land for the Sydney and Louisbourg railroad. That railroad was not the first in Glace Bay, in fact there were two small railroads, one operated from the old "Roost Pit" to the harbor, and the other from Caledonia mine to "Big Glace Bay" harbor. The first small railroad in Cape Breton ran from Sydney Mines to North Sydney shipping pier.

Before the turn of this century the Trans-Canada railroad ended on the mainland of Nova Scotia, near Point Tupper. People leaving Cape Breton travelled by horse-carriage or by ship. There were days when the train arrived at the end of the line at the Canso Strait; no transportation was available to Sydney; and a few walked nearly 100 miles. I asked my father

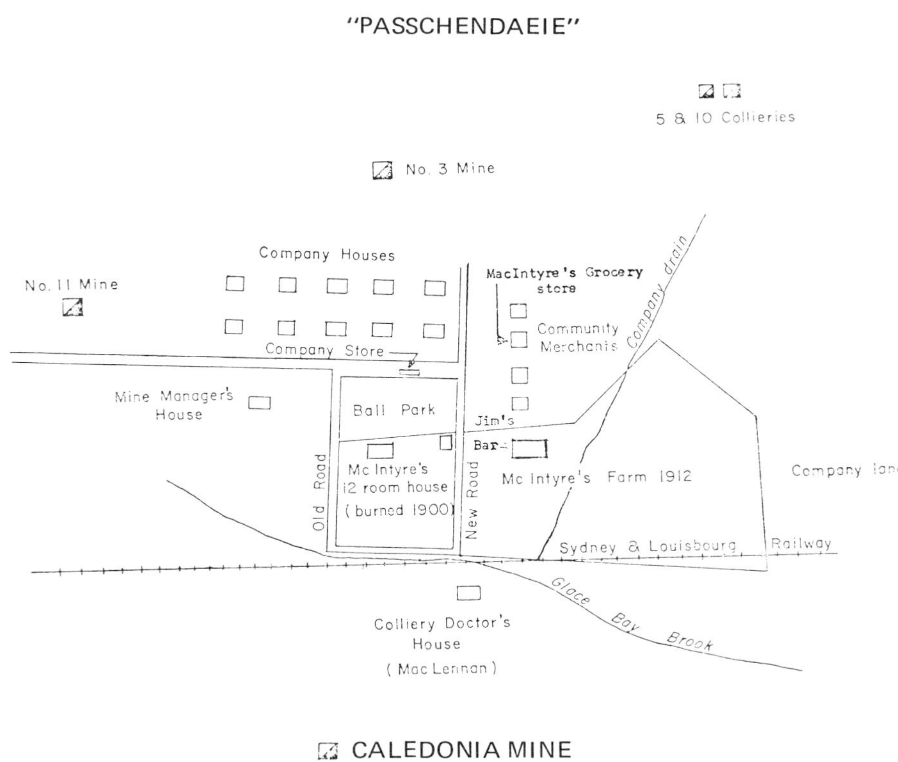
"PASSCHENDAEIE"
5 & 10 Collieries
No. 3 Mine
Company Houses
No. 11 Mine
MacIntyre's Grocery store
Community Merchants
Company drain
Company Store
Mine Manager's House
Ball Park
Jim's
Bar
Mc Intyre's 12 room house (burned 1900)
Old Road
New Road
Mc Intyre's Farm 1912
Company land
Sydney & Louisbourg Railway
Glace Bay Brook
Colliery Doctor's House (Mac Lennan)
CALEDONIA MINE

TOWN OF GLACE BAY POLICE FORCE – 1911

how he did it when he arrived back in Nova Scotia, and he replied, "I tied my suitcase to a post in the station, bought a bottle of rum for 25 cents and walked home to French Road."

There was no love lost between the coal company and my "old man" for many years, and the several houses that he built in Passchendaele were rent-free to the miners' families who were turned out of the company houses during the 1909 and 1910 coal strike. Sometime after the 1909 strike, the grocery store operated by my sister was closed in the Number 3 coal district, and in 1916 my father closed his liquor store. Then we started to farm on top of the "old workings" of the Caledonia mine. Chickens were imported from California, and their eggs hatched in the cold Spring weather, but the incubators and brooders were kept warm with old clothes and horse blankets. Some of the chickens which were hatched on our farm won prizes at Truro Agriculture College experiments, many years ago. Two beautiful farm horses, which were shown at "Fall fairs" in Maine, were owned by my brother-in-law, but they were stabled in our barns and earned "their keep" mowing hay and pulling potato diggers through land dotted with "pit holes". Hundreds of dollars were paid for animal-drawn farm machinery that bounced and was broken over rocks and tree roots.

Although my family was at loggerheads with the coal company over land subsidy, they never had a court case, and the company knew we took coal for our use from a "pit-hole" in the hay field for many years. And we allowed the company to drain "pit water" through the farm into the Caledonia brook. The drain became known in our family as the "company drain", and through a misunderstanding I lost several acres of land. The lease for the drain was given to my family by the coal company realtor, who told my father to move our fences in the cow pasture into company land, which we generally did during Spring fencing. Many years after that I visited the old land to look at my parents' graves, and I was told the new company would not renew the lease for the "drain"; the mines in that district were closed years ago. After a decade of farming around "pit holes and pit water", my father gave up after spending nearly all the money he made in other businesses, and we had an idea of going to Western Canada to raise sheep, but the old folks were reluctant to leave the land above the old workings of a coal mine. My father donated most of our farm for a graveyard and other church property; he never left, and my mother and father are buried in the soil where they were helped, by miners, to make hay and pick vegetables, when there wasn't any work in the mines.

Long periods of no-work-weeks in the Cape Breton mines

COMMERCIAL STREET 1904

just before the first World War and Glace Bay boom days were over. Hundreds of miners were standing on street corners, and the colliery pit whistles were silent. However, credit continued in the company stores for the people employed by the coal company, and another generation of colliers was experiencing the drought their fathers passed through in the last century. Miners who were buying goods from merchants outside the company stores were returning and purchasing their groceries at the "pluck-me"; small store owners could not carry large credit on their books and be supplied by the wholesale merchants in Sydney. The mines were in full production after 1915, but most of the young colliers were in France fighting a war, and many of them never returned to Nova Scotia; they are in the ground "over there". The mines were working daily, Winter and Summer, from 1915 until the end of 1921, but they were trapped in the slowing down of the Canadian economy during 1923, and the coal company reduced the colliers' wages 37½ percent without any board of conciliation. The wage reduction caused a coal production "slow-down" in the mines. The coal company could no longer pose as a giver of necessary staples to hungry miners. The public and the Nova Scotia government lashed out at the corporation when a top executive in the company muttered the detestable quote, "The miners cannot stand the gaff," during the 1925 strike.

The people were changing in the company rows; families needed more than just life's staples; large back-bills were increasing in the company store books, and the mine workers wanted more wages to pay their bills. The workers left the pits in 1923 and demanded something better than credit, when they were told, "The Nova Scotia coal miners had become one of the lowest paid colliers in the world." Nontheless, that statement was more of an assumption than a fact. I was told by a former Kentucky miner that he loaded coal underground in the United States during the great depression for 11 cents a ton. During the years after the depression the southern miners in the U.S. used a payroll credit system, but the stores were not owned nor operated by the many coal companies in each State.

The busiest hours of the week in the old company stores at Glace Bay were on Saturday morning; at least the most activity took place there at that time of the week, and it was obvious that the women and children were in no hurry to finish their shopping. Acquaintances spent hours inside the store talking about something or somebody connected with the company supervision, who was taken home in a wheelbarrow last Saturday night. The children had more fun "waiting to be waited on" than most kids today would have at a circus. The boys

DOMINION COAL COMPANY
GENERAL OFFICE

GLACE BAY
1901

would run through the store "lickedy-split", hanging on to large kerosene cans which were later filled with oil and the spouts plugged with small potatoes. The kerosene oil was used for lighting lamps in the homes, but before this period of time at the collieries the miners paid for the oil that they burned in pit lamps at work. The kids played tricks on each other, and vexed the store clerks with screams from the girls as the boys pulled on the long "pig-tails" of hair on their heads. The girls generally carried the molasses cans, and called the boys names as they played "ring around the rosey", around the pot-bellied stove in the center of the dry goods department.

Most of the boys had nicknames, and the pseudonym of "Billy-bob-tail" or "Joe-pluck-me" might be called out to a boy by a clerk back of the grocery counter to get out of the store because he concealed an apple under his jacket. During this confusion on the floors the clerks shouted, "who's next?" but their customers were listening to the loud voices coming from the store office, "someone was demanding a grocery order." There were cases when the mine manager interceded at the pluck-me for people who were off work for long periods of time for reasons unexplained, "and for others who were trying to explain excuses."

There were days in the old Cape Breton company stores when a Highland collier arrived there "madder then a wet hen", and in a voice which was loud and a clear mixture of English and Gaelic curse words, accused the store clerks of using super salesmanship with his children, who were buying almost all the articles displayed on the shelves of the store. He shouted, "I hope all of you burn in hell. God-damn-it, you have taken most of the wages that I have earned since last Christmas, and I never had a drink of 'the stuff' since Easter." The company store was taking most of the money he made working at datal time, and every pay sheet was a "bobtail" and some weekends with a bottom line of zero. When the enraged Scot demanded an itemized account of his family purchases it was made known to all in the store that he could expect more "bobtail" paydays before the Summer passed. The defeated collier returned home and reprimanded his family for over-buying at the "pluck-me". There were cases, after discussions of extravagant buying in the household, that a new set of dishes was bought on Monday morning at the company store. The coal face miners were paid tonnage rates for the coal that they produced, and they generally made a dollar or two more than the datal time worker. However, the contract colliers' earnings would be reduced for long periods, waiting for "places in the pits on low datal wages", which reduced the miners' purchasing power at the store and caused people to acquire back accounts that were paid at a later date.

Many of the women in the mining districts of Nova Scotia made clothes for themselves and their children from bolts of cloth purchased in the company store, and also knitted socks that the men wore in the mines. There generally were ladies in each district who were accomplished dressmakers, and the girls would have new dresses for class graduations. The boys in grade school wore white shirts on the day the class picture was taken, but some refused to part with an old sweater with a large safety pin in the neck.

The company store employees were colliery people, men and women, trained for jobs in stores and offices, and everybody knew they were company confidants who worked for their livelihood in trusted positions. The store clerks were honest and capable, nonetheless, there were persons who shunned the company stores simply because clerks shamed people in public over credit. There were other people who had a fear of the stores' system of selling merchandise on a workers' payroll number where they worked.

The company stores went out of existence in Cape Breton mining districts in the year 1925. The stores were looted by young colliers that Summer. After, it was rumored the company stores were soon going to close permanently, and a few days after the looting, the stores in Glace Bay were burned to the ground. It became known after the company stores were raided that the general merchandise in the buildings was below the regular stock carried in each store, and that the clerks had prepared for the raiders. Shoes of mixed sizes and colors were placed in boxes; the grocery departments were found bare by the looters. Practically every newspaper in Canada told the story. Glace Bay coal miners burned and looted company stores, and they were sure the miners would have many hungry days in the future.

Sure, it is a fact that young men and boys from the mining districts looted the company stores during 1925, but it may not be true that they set fire to the buildings that housed the old "pluck-me" stores. Some years ago this writer published a short article which was headlined "The Pluck-Me Store", and while doing research on that story a former fire inspector told me that he was the little old man with the false gray beard, on the two-wheel bike, who was seen around the stores before the fire alarms were sounded. The store at the "Mines" was not burned at that time, nor was the building at Dominion Number 6.

The looting of company stores in Cape Breton during 1925 was not the first time that happened in North America. My father was in a small mining village in Pennsylvania, during 1875, when the coal company store was sacked and burned by the "Molly Maguires". Most everything in the store was thrown out on the street before the building was burned, but most of the colliers did not touch the plundered merchandise. The miners were Irish and Scotch immigrants, who knew the company police would be searching their homes in the morning. Reprisals for stealing company goods at that time was jail, and the family was told to "move on".

The "Molly Maguires" was a small group of aggressive miners in the union, which was known as "The Workmen's

Benevolent Association", formed some years before the U.M.W. During these days a labor leader, Jack Kehoe, was hung for the murder of a company official, and 100 years later Jack Kehoe was pardoned by the State of Pennsylvania because another man committed that crime.

During 1919, the U.M.W. returned to Nova Scotia, and the colliers in District 26 started to pay the money which the union spent during the 1909 strike. The return of the American miners' union was peaceful, no quarrel with the company or the provincial union functioning in Cape Breton during the first World War years. Some of the older union men wondered how it was this way after its bitter rejection during 1910. The P.W.A., now forgotten, became a company thing, through the intermingling of individuals looking for security for themselves at the expense of other people in the old "Lodge". Not all — Neil J. Gillis, secretary of the Provincial Workers Association of Nova Scotia, was responsible for a "Bill" in the Nova Scotia Legislature which made eight hours a shift in Glace Bay.

During cases of severe injury and death in the colliers' families, the company officials would be lenient in many of these circumstances, and money which was urgently needed was advanced to the miners by the paymaster on the advice of the colliery manager. Money was also loaned to employees to move their families into the mining districts, and some of the European immigrant miners had their transportation paid, which was later deducted from their wages. At times small advances on the next pay could be "worked" at the office if the mine manager was having a "good day" underground, but the note had to be signed by the chief colliery clerk before it was paid to the applicant at the company head office in Glace Bay. Many regulations used by corporations are supposed to be foolproof, but workers often find ways to beat them, and I guess trying to get a few dollars in the middle of the week came under that category. One of these approaches on a sympathetic mine manager was to claim that an uncle was dead up in the country, and the applicant, notwithstanding that his relative was buried many years before that idle day at Number 2, was close to tears when he asked for a few dollars to attend the "poor man's" funeral.

Sure, it is true some people made a little noise on the weekends, and had their "lamps stopped", mostly because they did not report for work on Monday morning. Chronic absentees on Mondays were generally told to take a few more days off in the sun before coming back, and asked the manager for their lamps to make another start underground. When the miner had an advance to pay the colliery chief clerk after a week off work,

This picture was taken after the last mine in Reserve was inspected and later closed. Most of the miners were hired at other collieries.

he was generally told, "be on the pit bottom in the morning or don't ever come back." It was a general custom in Cape Breton mining districts to stop the miners' lamps when they were off work for two or three days. The lamp house attendants would not give out any lamps with a ticket on them until they were cleared by the mine manager. Some people were asked to take a liquor pledge from their parish pastor, and come back at another date for a slip to get their lamps. Chronic absentees told that they joined one of the temperance societies which became popular during the years of the first World War. The heavy drinkers who lost work through liquor took their non-alcoholic pledges for six months, but there were others who rejoined the L.O.C. every second Sunday when paydays were semi-weekly at the mines. However, to end this discussion on the uselessness of "booze", it's a fact that a great majority of the miners seldom lost work without a good cause.

Several of the liquor vendors in Glace Bay closed their shops after the "Scott Act" was abolished during 1917, but many remained open, although they were often harassed by the new provincial police. The Nova Scotia Temperance Act was repealed during the 1930s, and the government took over the liquor business, and the "old bars" closed forever, all over Nova Scotia.

During the periods that the N.S.T.A. inspectors searched "barrooms and fishermen's hides", "rum row" on the Eastern Atlantic Coast became inshore fishermen's stock and trade, for a few prosperous years. There never was a shortage of rum in Halifax or Glace Bay, at that time. Rum sold at two dollars a bottle, and the railroads shiped beer into Cape Breton towns for several years after the old Scott Act passed away in Nova Scotia.

The mining town of Dominion, which is adjacent to Glace Bay, had at one time many Italian families that arrived in Cape Breton early in this century. During the time that the Nova Scotia temperance inspectors tried to enforce the liquor laws, the Italian miners purchased beer from Quebec cities by railroad "box-cars". That draft beer kept these hard-working people happy, and on the job any day of the week when needed. During underground emergencies at the colliery on the weekends, when large falls of stone had to be cleared from the mine haulage for the mine to work the next day, the Italians were available for the job. The mine overmen searched for men during these unforeseen occurrences, and generally found them, or most of the people needed in "bug row" on a Sunday afternoon. "Bug row" was a beer drinking establishment in the town of Dominion many years ago.

Most of the lines that I have written in the last two

paragraphs I have experienced and seen with my own eyes, and it must be admitted the coal face collier is not only skilled at his work, but a very independent laborer, and you do not push him around. During the years which Cape Breton mines worked six days a week, all day shift, many of the contract colliers called in a week on Friday. Friday's coal tonnage was the highest in the week, and Saturday's the lowest, but some managers held back a part of Friday's coal and supplemented it to Saturday's output. Saturday was pay day in Cape Breton for many years, and Glace Bay was the largest town in Canada, but the "Bay" was never the kind of mining town that Robert W. Service has written ballads about. Nonetheless, some young news reporters did their best on paper to make our town look like "hell" during the few general coal strikes in Nova Scotia.

During the Summer of 1977, I visited one of the three largest collieries now operating in Cape Breton mining districts. There are approximately 1,000 people employed at 26 mine at Glace Bay, and most of them are underground workers. Noticing lunch cans around the surface, I was told by a surface collier that many of the coal face miners do not take their "piece cans" underground, because they think they are over-loaded with a gas mask and an electric lamp. Later in this paper the writer will describe the number of articles the early Caledonia miners carried as they walked long distances underground into the mine workings.

The Cape Breton miner has the same model of lighting as the best equipped North American miner used underground, although the battery is a little over-size. The lamp is strapped to one side of the worker's body at his hip, also held in place by a belt is a gas mask. The electric lamp should burn bright for approximately 10 or 12 hours, and the gas mask has roughly one hour of air to be used in emergency. In the large Cape Breton mines very few workers carry tools during their transportation from the pit bottom into the coal faces; it was an unsafe practice, and men were often injured by sharp tools in the "riding rakes". The Number 26 miners are transported from the coal faces through cold airways, and each collier wears a heavy jacket for warmth.

The most loaded-down person in the pits today is the "shotfirer", who at one time held an official status in Cape Breton mines. The former official blasts coal and stone from the mine strata after he has examined the immediate atmosphere. He carries a flame safety lamp and a leather case filled with detonators that are used to explode mine drillings after they are loaded with explosives. That is only a part of the "shot-lighter's" burden, as he also has a gas mask, a small gauge electric cable over 100 feet in length, a heavy rechargeable battery, "maybe", another lamp, electric, to see and work in a safe manner. It is known that along coal faces where blasting is required others are making a hurried examination of drilled mining, because the shot-lighter cannot see to do this work with the lamp he uses to examine the atmosphere. The result of missing a "crack" and the depth of the mining has produced "flame-shots" which will explode a dusty environment which cannot be detected with electric gas testers. This statement may not be true or a general habit, but this writer was told that the only people who carry the flame safety lamps underground at 26 colliery are the mine examiners and the shotlighters. I was also told that coal face miners do not have their gas masks on their person along the faces, and are hidden at the top of the wall material road, and few would be found in an emergency. There are thousands of colliery workers down in mines all over the world who will never be given a gas mask for their safety in the pits. The few colliers, who are supplied with the masks in gassy coal seams, should have them ready to place over their faces during a "gas expulsion", even though this may never be necessary. It is the responsibility of the workers' supervisor to see that his men are working in a safe manner underground. This writer worked in mines and many gassy places for 33 years, *"no coal company pension"*, and I have only seen one large gas expulsion in all these years. I have inquired from older mining people at that time about this abnormal gas release; they never saw nor heard of such a happening in Cape Breton mines.

Passchendaele is a small community in the town of Glace Bay. Although the colliers were a mixture of Irish and Scotch descent, the residents of the ward did not have any feeling of separation over religious beliefs. The Catholic residents, with the help of others in the village, constructed a church and a school in 1903. "Nuns" taught children of several religious denominations from Caledonia and Dominion Number 4. The first Catholic priest at Saint Anthony's parish in Passchendaele was Reverend Ronald H. MacDougal, who was loved by all the people in the district, and many of the children born in Glace Bay were named "Ronald H.".

Passchendaele had several pseudonyms since "Jim" MacIntyre, during the last century, cut a road through a surveyor's trail across the Caledonia Brook and built the first house in the hamlet which now has the "nom de plume" of Passchendaele in the town of Glace Bay.

Passchendaele had several "non de guerre's" over the years since my father arrived there with his family. Two mines were started around the turn of this century, Numbers 4 and 3, and

"PASSCHENDAELE" MANY YEARS AGO.

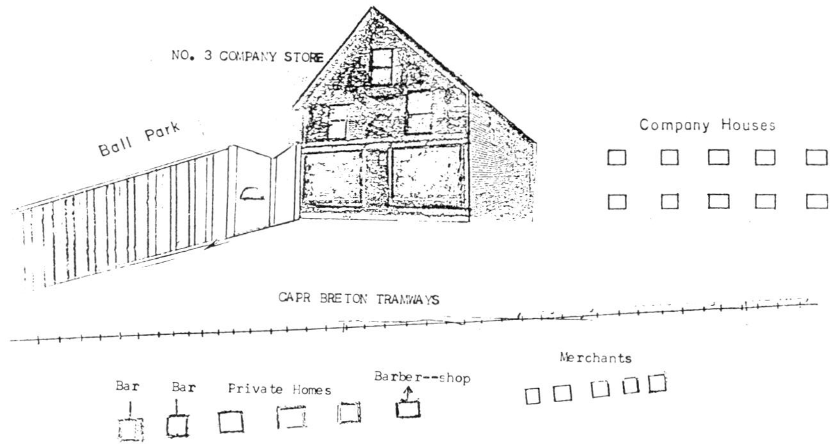

During the 1909 strike the coal company took over the Passchendaele ball park, and when the soldiers from Halifax arrived in Glace Bay they set up camps and mess kitchens in the center of the field. The soldiers shot one man, accidentally, an old shoemaker that lived in a house owned by my father at the edge playing grounds.

the village became known as Number 4; later 4 was stopped and Number 3 was worked for nearly 15 years on the Phalen coal seam. The community then became known as Number 3 for more than a decade, and many foreign-born miners arrived there, mostly Italians who moved to Dominion and Donkin when Number 3 closed down. Two or three years before the first World War started in Europe, old Number 4 was started once more as Number 11, as Caledonia took "four" as a number. Number 11 mine worked for many years; it was the last coal pit in that district, and when the second war was over the colliers who returned from France named Number 11, "Passchendaele".

Passchendaele was a thriving community, at one time, which served the people in the district center and the surrounding villages. Two mines were operating at Birch Grove, 21 and 22, and two in Dominion Number 4, 3 and 11. Most of the residents from across the brook, Lake Road and Maple Avenue, and the settlement which accumulated near the Marconi Towers, bought their weekend supplies in Number 11, many years ago. There were two liquor vendors and seven general stores, besides the company store. There was a professional baseball team in the old Number 3 district before I started school, but rugby became the most popular sport in Passchendaele. Two of Canada's best boxers during this period, Mickey MacIntyre and Roddie MacDonald, were neighbors in old Number 4. Passchendaele had electric street car service when cities in Ontario were using "horse cars" for transportation. The service was from Sydney to Passchendaele until the "iron bridge" was constructed across the Caledonia Brook, then the line moved into the business streets of Glace Bay, and through New Aberdeen, Bridgeport, Dominion and Reserve. The iron bridge was a lonely spot at night, and it was known as the "haunted trestle" that prompted one of Gregor Farrel's sons to write a book headlined "The Ghost of the Iron Bridge". I have seen several hundred people there at night after somebody reported a good tale or that he had seen the ghost come out from under the bridge. A few families were moving out of the village when it was learned old Number 3 was going to close, but it was told that one family which moved to Number 6 (Donkin), left because their father met a woman near the bridge at five o'clock in the morning on his way to work in the mine. Marconi was in Glace Bay at that time and he was interested in the iron bridge ghost.

Card games were popular in Passchendaele 50 years ago, and teams of four people played 45s, and challenged other teams in the mining districts. During the Winter tournaments many articles and animals were played for in the halls and homes;

"PASSCHENDAELE"
"The Front Row Houses" - 1914

CAPE BRETON TRAMWAYS ON INCLINE AT NO. 11 – 1914

generally the prize was a horse, a side of beef, "spring pigs", fighting dogs or roosters. A few of the dogs and the roosters were bred in Passchendaele, although many were imported from the United States. There may be a rooster "pit" around today, but happily, fighting dogs were forgotten in Cape Breton many years ago, and "bingo" has replaced "45s".

The first ballfield in the Glace Bay mining district was the Number 11 football park, but it was first constructed for baseball in the year 1900. The Passchendaele rugby squads never made the Nova Scotia senior league playdowns until the '40s when they won the McCurdy Cup. Number 11's great rugby teams were in the '20s and '30s, but their opponents in the Cape Breton playoffs were Caledonia, their "arch-rival", considered the best rugby team ever formed in Canada. People today who remember the Caledonians on the rugby fields in Eastern Canada, claim they were capable of beating a few of the rugby teams from Europe that tour North America every four years.

RUGBY
TEAM
DAN McPHERSON EX.
MALCOM CAMPBELL EX.
H. JOHNSTON
W. JOHNSTON FORWARD
A. McKINNON EX.
W. McDONALD
MANAGER
J. FLYNN EX.
W. SCOTT FORWARD
R. McINTYRE HALF
B. FERGUSON FORWARD
J. McRURY
INTERMEDIATE CHAMPIONS
1924
HALF
A. CAMPBELL

Since that period I have watched most of the great rugby teams play from England, France and New Zealand, and the game is played differently and much faster today, compared to rugby squads of past decades. Caledonia's greatest opponents were in Nova Scotia rugby leagues and playoffs; they lost the Eastern Canada rugby titles several times during playdowns with the Halifax Wanderers Rugby Club, before the club hired a professional coach, a former international rugby player, "John McCarthy".

The rivalry between the Caledonia and Passchendaele rugby squads in the 1930s was only equalled when the Glace Bay Legionnaires entered a team in the Cape Breton Rugby League, under the banner of "The Legion Guards". The Legion team was formed of boys from the Glace Bay High School, and others from Caledonia, Number 11, Donkin, and one or two players from the Halifax Wanderers' Club. The writer was captain of the "Guards" for a couple of seasons, but that team was disbanded after several players returned to Caledonia and Number 11. Passchendaele's rugby squads were at their best during the '20s and '30s, and it is stated that the 1924 team, which only lost one game in three years, was one of their best rugby clubs. However, for some unknown reason, that team never played in the Cape Breton senior league, although it was capable, and may, in all probability, have represented Cape Breton in the Nova Scotia playdowns at that time. The Passchendaele rugby team of the 1924 era was broken up because of the 1925 Summer coal strike, and that squad was never to play again as a unit, as many of the boys were working that Fall in the United States. Some years ago when I was working in Indianapolis, Indiana, I was asked in a letter from Glace Bay if I had a picture of our 1924 team to send to Johnny Butts, then an M.D. and surgeon in New York City.

The Caledonia team defeated most of the teams in the provinces of Quebec, New Brunswick and Nova Scotia, but could never get a game with a rugby team from British Columbia. British Columbia has the best rugby team in Canada today, but they will have to have a lot more practice to achieve the passing skill of the "old Caledonians".

The rugby rivalry between "Cal and Eleven" remains among the older inhabitants in both districts today, and former games are rehashed after pension cheques are cashed and a few of the "old boys" take off for the Legion with a few dollars to spend. However, there are only a few survivors of the photo shown here of the 1924 Passchendaele rugby team, although "Yap", "Hubert", "the Buck and the Bear", had several drinks together the night in 1977 when the Legion branch in town held a

CALEDONIA ENGLISH RUGBY TEAM

EASTERN CANADIAN CHAMPIONS – 1929

Front Row: Left to Right, Graham Crocker, Bob Jackson, Mascot Chas. McVicar, Broph MacDonald, Geo. Nicholson. Second Row: Left to Right, Manager Chas. McVicar, Melvin Sheppard, Dan Nicholson, Geo. MacNeil, Ewan Hillier. Third Row: Left to Right, Tom Jackson, Jim Driscoll, "Toots" Boutilier, Jim Leslie, Louis Morris, Clarence MacLean, Sandy MacMullin. Back Row: Left to Right, Wm. Wilton, Stan Scott, "Hun" Lawley, Neil Nicholson, "Jut" Wadman, Treasurer Jas Ferguson.

banquet for former Cape Breton rugby players. "Sander" was there for a short time only as he drank his quota many years ago with me and "Duffy". The Caledonia boys in their picture here and later years have "thinned out", and several have died in the mines and World War II. There were several of the "old Guards" at the Legion banquet; that team also had casualties in the last war and the mines. Art Sharp, "scrum half", died in the mine. The recent explosion in Number 26 colliery took the life of another person named "Sharp", Art's son. Donkin, Number 6, was also represented at the Legion the night of the banquet, and it was a happy reunion, as "Scrogy", Gordon and Clark were with the Legion squad during the years I played with the same team.

The picture shown here with the young boys in football uniforms was taken during the 1930s. They are miners' children from Passchendaele, in the coal town of Glace Bay; two or three

" THE NO. II INVADERS "

Some of the players identified were (seated in front): Arthur Chiasson, Burns and Billy McMullen; (second row): John McNeil, Arthur McLeod, Harold Johnston, Joseph Pilling, Father Andy Hogan, Fred Campbell, ? O'Keefe and Red McMullen; (standing): ? Burns, ? Donahue, Ralph McNeil, ?, Francis Currie, Rannie McIntosh, Felix Chiasson, ? and ? McIntyre. In back are the coaches, former Councillor Bill Cann and Philip Campbell.

decades before this period these little boys would be wearing "pit clothes". One of these lads named Hogan, became a Catholic priest and is now a member of the Federal House of Commons in Ottawa. Father Hogan has made a noted study of economic science in regards to practical affairs of workers and communities. "Andy's" father was a mine boss at Number 11 colliery when I was playing rugby in the Cape Breton League. Rugby is no longer played in the colliery districts of Nova Scotia, nor did it return to its former greatness after the second World War. There were several reasons for the fizzling of rugby after such a promising sport in Cape Breton. Although only two teams were left in the league, one good reason was the lack of publicity in the daily newspapers, and the radios. Donald MacInnis, a former rugger, and a retired member of the Federal Government, is making an effort to bring the game back to Glace Bay, but he must get help from organized clubs too. Many young men are just standing around the streets after work looking at themselves and talking about baseball. "Donald's"

SOUTH STREET BALL PARK

father was killed in the Caledonia mine during 1920.

However, the game of rugby is returning to the mainland of Nova Scotia and Newfoundland, but in Cape Breton, a natural spawning ground for rugby players, the public expects the new players to be the "warriors of old", and it may be a few more years before the game will start from scratch and be played on the grassy fields at Passchendaele and South Street parks.

The game of rugby was developed from a former English school for boys. Rugby College, in the year 1823, played a game which became known as rugby, and players were called the "Ruggers". Since that period rugby has been played in leagues all over the world. American and Canadian football was once played under revised rugby rules, and the first international intercollegiate game of football played in North America under these rules took place at Cambridge, Mass., between Harvard and McGill University students in 1874. Most all the large colleges in Canada and the United States have rugby teams. Soon after I retired I watched several games in Boston, which were sponsored by the Harvard Rugby Club. Three of the teams were from England, France and Ireland. These teams carried "underage" players, who filled in when the regulars were resting. The game between England and France was the highlight of the games, which England won. But not before the fans gave the French fullback a standing ovation. That youngster, only 18 years of age, picked up the ball near his own goal line, broke away from tackles, and sped down the field. At the 50-yard line he was hit by the English fullback, but not before he dropped the ball on the ground and kicked it over his opponent's goalposts and delayed the game for about ten minutes.

Rivalry in games between Scotland and England is a part of the "rugger" character that encourages healthful fitness. Nonetheless, competition may be carried beyond the realm of sportsmanship. Although England and Scotland never went to war over rugby games, the heat during national playdowns was terrific. The Scots wanted the name of the game changed to "Scotish Rugby" after they won the playoffs two years in succession, and later when they lost it became an English game they played against mostly Welshmen. The friendly competition which excited the Cape Breton rugby fans in the Colliery League was there until the game died 30 years ago. I have many fond memories of the days when "Shep" was the Caledonia fullback, "Gyp" and "Jam" in the Number 11 backfield, and Johnny Vey the stand-off half for the Legion Guards. Halifax rugby fans gather at Keith's Hall once or twice a year, and talk about the

"Sad Sam" Jones played in the outfield and pitched for the Passchendaele professional baseball team at one time. Fred Lewis was manager of the team.

World champion Boston Red Sox pitching staff. (Left to Right) Sad Sam Jones, Carl Mays, Babe Ruth, and an unidentified fourth player at opening of the 1915 World Series. Fenway Park, September 9, 1915. 1915.

During Sam's stay in Passchendaele he lived with Mr. and Mrs. Arthur MacLennan, the club's business manager. Later the MacLennans moved to Sydney, and Mr. MacLennan became one of that city's aldermen. ermen.

years Ritche McCoy and the Nickersons rugby skill, when "Alex" was a young newsman.

Baseball was played in the mining towns of Nova Scotia before rugby, and the fields, which were cleared of trees and stones at the turn of the 1900s, were made into baseball parks. The two coal towns of Springhill and Dominion were baseball towns and never played in organized rugby leagues, but Reserve had the most noted rugby squad in Cape Breton at one time, 1908 or 1910. The rugby team from Reserve Mines was a "semi-pro squad" with known professionals who were paid to play the game on Saturday afternoons. Several of the ruggers were college players, and three or four English miners who played rugby in the "old country". "Billy Reid" was a member of that team; he was considered the greatest rugby player ever to play in the colliery league. "Billy" was an international and professional rugby player in England before he arrived in Glace Bay, and it was stated that people came to the games to see him play and use the "dummy pass". Tommy Jackson, a Caledonia rugby player, acquired the nickname "Dummy" for his skill with the fake pass.

"Baseball" is the miners' greatest love in sports all over North America. They play the game, watch it on TV, and crowd into their local ballfields during the Summer months to see teenagers and juniors play. I have met miners from the American coal states that never miss the baseball World Series. Several of the greatest ball players in the "big leagues" learned the game in coal towns. Today, however, the majority of the players are college people, who sell themselves to the highest bidder.Nova Scotia miners who work under the Atlantic Ocean for their livelihood had to be informed after each inning the score of a World Series game in New York at one time. The messages were sent over the telephone into the underground dispatcher's cabin, and then relayed to miners on a "longwall" coal face under the deep sea coast of Cape Breton Island. The mine workers were given assurance, after a lot of discussion on the surface, that most of the information on that particular game would be relayed to them. The company paid an office clerk a few extra hours, and those miners "hand-loaded" 600 tons of coal on the three to eleven shift. What motivated these people? Simple wants. Human beings need more than food and shelter, and corporations know they are dealing with people whose thoughts are not theirs, but many of them fail to understand why they should have relationships with their own employees. It is no secret that management and union officials are at each other's throats for years because labor and capital fear each other.

Passchendaele's last championship baseball team defeated every team in junior league play-offs, and a few clubs that were senior teams. This team was maded up from boys that lived all over Glace Bay.

Glace Bay No. 11 Antonians 1952 Maritime Junior Baseball Champions. Shown in the front row are: (left to right) Naish Batten, Cecil Brufatto, Tommy Paul MacNeil, Eddie Michalik, Bobby Wilton, Roy Sulliman, Rod "Yogi" Boutilier. Back row: Buddy Bonnar, Steve Hogan, Jackie "Lily" MacLellan, George Nicholson, "Big" Jim MacLellan, Coach Jim O'Dell. Missing when photo was taken: Jim "Lefty" MacVicar, and Jerry MacLeod, Manager.

Photo By Abbass (Courtesy of Rod "Yogi" Boutilier)

Baseball was played professionally in Cape Breton at the turn of this century, and the community now known as Passchendaele had a team in the league. Most of the players were miners, but a few were imports from the mainland of Nova Scotia and the United States. "Sad Sam" Jones was one of the Boston Red Sox pitchers that won the American league during 1915 and 18. The other pitchers who helped "Sam" were also "greats" — Babe Ruth and Carl Mays. Maybe a year after the 1909 coal strike in Nova Scotia, Sam Jones pitched for the Passchendaele ball club, then known as the Number 3 "bear-cats". Several of the colliers who made their home in Number 11 arrived there through baseball abilities, and they came from mining towns in Nova Scotia. During the Winter hot stove league in Cape Breton, it was told and talked about that "Bobby Burglar" Fergson, a miner's son, was the best "home brew" to play in the old professional league, and that he was about to be signed by the Detroit Tigers when he enlisted in the first World War. "Burglar" was wounded coming home from the war, and his baseball days were over. "Burglar" became a mine mechanic, and he was old and slow when he had a fatal accident at IB mine.

The Cape Breton professional baseball league was disrupted several times over "short work weeks" and a few strikes, and no "pro ball" was played in the colliery towns since before the second World War. Passchendaele had a "Junior team" after the war which became the Nova Scotia champions, and their star pitcher, Wilton, was given a "try-out" with the "Braves"; Bobby is now teaching school in California. The game was not always played in leagues, and teams from Sydney and the colliery towns challenged each other for games. Most of the young boys who learned baseball in their teenage years, between the two great wars, did not play the game after they started to work in the mines. It was different with football. There was a series of games to be played between boys from Sydney and Glace Bay who were in their late teens, which was talked about for years. The Hebrew "All Stars" from Sydney challenged the Glace Bay "All Stars" to a series of games which was never completed because of an aged boy named Horace Wells, a star junior who had not played ball that Summer, but the "Bay" team were going to play him, and the series ended there with the "Hebrew Stars" leading by one game. The boy mentioned, Horace Wells, was fatally gassed in IB mine some years later. Baseball in Cape Breton today is mostly "Little League", and a team from Glace Bay lost the Canadian National Championship in Western Canada in one extra inning in 1976. Michael Meechan, has done a splendid job with the youngsters, and "Mike" was named "Man of the Year" for that, and other little things that visitors notice in the make-up of retired miners in Nova Scotia.

Rivalries at the collieries did not end in sports; coal outputs and popular mine officials were discussed when the collieries talked about their work and the mines where they worked. Records of cost of coal produced from the Cape Breton mines in the year of the '20s and '30s of this century were the top secrets of mining executives, but there are a few people left in the colliery towns who remember. Tommy Casey was the youngest person to be appointed a mine manager in Glace Bay, and older officials did say he was too young to be a colliery manager. Casey was a popular underground official with the "run-o-mine" workers when he was placed manager at the Number 11 mine in Passchendaele. Mr. Casey became manager of Number 1B colliery during 1928, and during 1930 coal which was loaded by hand in underground "rooms" was placed in coal "hoppers" under the "bank-head" for $3.35 a ton. That cost was made through hard work by 1B miners, officials, and the respectable relationship the people from New Aberdeen, Bridgeport and Dominion had for their mine manager at that time.

Production and cost promoted and demoted mine officials,

and each underground overman had his own organization. The coordination of the Boss's people would make or break him as he was responsible for the safety of the miners in his section, and the cost of the coal sent to the pit-head. When a fatality occurred in a mine, it was the overman who was questioned by the Nova Scotia Mines Department at the investigation, but he was accompanied by the colliery manager and a company lawyer.

For the many years that Cape Breton miners worked on contract, the mine overmen were responsible for the money paid to their workers, datal and contract, and they bargained on jobs and payments, but did not always agree with each other.

The Glace Bay brook which flows through Passchendaele has the traits of a large river, and has overflowed during which two large bridges were washed out by water currents during Spring floods. The coal company constructed dams across the brook from Caledonia to Reserve Mines, which held water for the colliery steam boilers. The early colliery residents in Passchendaele washed clothes at pools in the brook before the water was piped into their houses. The brook was considered company property by the people who lived in Caledonia and Number 11, and watched by the company police. During the Winter months the public was allowed to skate on the dams after the water froze, but in the Summer the dams were "off limits" for swimming, and young boys who swam in the brook generally found a deep pool not watched by the police. However, the dams were alluring and deep, and the writer, with several small school chums, had their clothes taken to the colliery office by the coal company police, where the mine manager held court over five naked boys looking for their trousers. After a scolding we were allowed to go home and often punished by our parents for swimming in the deep water dams, where several boys were drowned. Three boys from Brookside Street were lost through the ice on the Caledonia dam, one on the Number 4 dam, and two little boys drowned while swimming in the Reserve water dam. There were several near tragedies at the dams that were never talked about after the boys returned home.

The mines are now closed in Passchendaele, although relics of buildings remain in the old pit yard that remind people unemployed that their fathers once had jobs in Number 11. Some come and visit the places they left many years ago, and search for old friends that most of the residents there today never knew. The old names are common, and Pastor MacIntosh operates his father's store that was first opened where the Saint

The source of the Glace Bay Brook is back of the Sydney Airport; a small lake which was at one time abundant with large trout. The brook flows through the village of Reserve outskirts, Passchendaele, Caledonia, and into the Glace Bay Harbor.

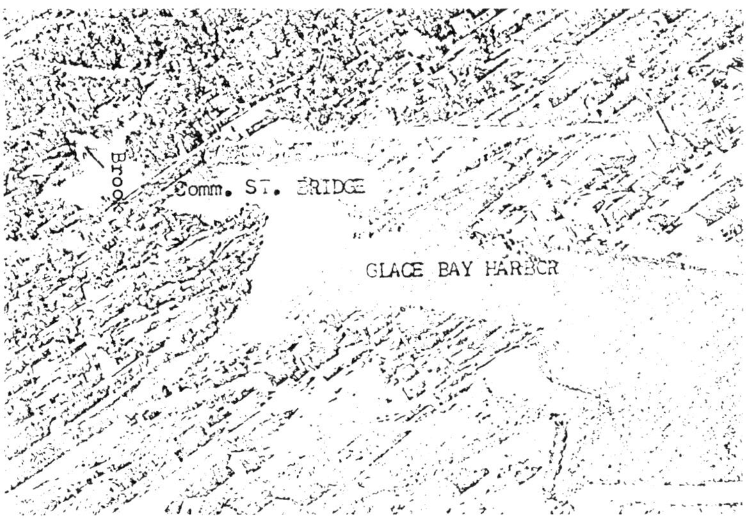

Anthony school now stands. The ball park is empty at Passchendaele, no football posts, but maybe there is a baseball diamond covered with dust and grass. The field is fenced and clean, just waiting for the shrill sound of a rugby whistle or an umpire to shout "play ball". Most of the company houses are now owned by the residents, and from what I saw there last Summer they were in good shape inside and out, and a few of the "old-timers" had flower gardens. "MacIntyre's" farm now contains a church, a school, a convent, and several houses and a graveyard. The house where this writer was born and raised is now the home of the parish priest in Passchendaele.

The old brook is quiet now, minus dams and boys playing hookie from school, as it flows gently under the "iron bridge", which was considered the dividing line between Caledonia and Number 11. In the next few paragraphs I'm going to cross over that old bridge into Caledonia, where I visited last year, looking for "Jimmie Dannie Ann". I became lost after we passed the top of "old English Town rows", and I was reading a large billboard which stated "Caledonia Heights". Later I met one of the "Oliver's" who directed me out of the old colliery yard.

Most of the first families in Caledonia were Highland Scotch people from the Mira District, which covered most of the settlements of the County of Cape Breton. There were a few

CALEDONIA MacINTYRE PIPE BAND — 1902

DAN MCINTYRE, CHARLES JAMIESON, JOSEPH MCINTYRE, MICHAEL MCINTYRE, JOHN JAMIESON, PETER MCINTYRE, JOE MCINTYRE, ARCHIE MCINTYRE.

French people who were in Nova Scotia since Louisbourg was a military fortress, and several Irish families that arrived in Glace Bay in the middle of the last century. The Polish families were in Caledonia and working in the mine during the first years of the 1900s. Most of the "Poles" attended the Catholic church and school in Passchendaele with me sixty years ago. The people from the Mira settlements had their own homes in the "Country", made fishing boats, and raised a few sheep and cattle. They also weaved wool into cloth and made clothes for themselves and others; many of them were nicknamed "the Tailors". These Scotch Highlanders never intended to spend a lifetime in the coal pits. They came to the mines to make a few dollars through which they would upgrade their land and cattle. Nonetheless, only a few returned to the land their fathers cleared and worked during the early 1800s. Mining became a way of life for the people in the town of Glace Bay, and it cannot be denied that the Nova Scotia colliers became skilled miners by hard work and study in the colliery night schools.

The Caledonia Colliery became the pride of the Dominion Coal Company, after Alex (The Pond) MacDonald was appointed manager, and Jack Casey became the underground manager, during 1902. The MacDonalds and the Caseys are a part of Cape Breton's mining history, and it was Jack Casey and a man named Wilton who taught the first colliery mining class in Caledonia. The merits of the Caledonia miners and officials were proclaimed in mines all over Canada. It took a lot of hard work and study for the Caledonia mine to become known as a "clean coal mine". The first workings were gassy, and the mine horsemen, "drivers", lighted small gas feeders along the haulage roads. Coal pillars, which are left in mines to protect mine shafts and the underground roadways, were never larger than 12′ by 12′ blocks of coal in the first workings at Caledonia, and some of these were crushed when the east deep was driven through the old workings of the west deep. Why did these early miners leave tiny blocks of coal not more than 12 feet square? It took years to remedy the roof conditions on the main haulage from number 5 to the pit bottom, and at one time it was thought the mine would be lost. Because air in the early mine workings could not be pushed into the coal faces with pillars 50 feet in length; the coal pillars were shortened and "cross-cuts" were driven every 12 feet to establish an airway. The rooms were 16 feet wide and six feet high, but the air pressure was about nil after approximately 25 feet from the air cross-cut. The ventilation was not moved into the faces with brattice, "there was not any", and the first brattice fabric ever used in the workings of a mine in Nova Scotia was at the Caledonia colliery, after Jack Casey was appointed the colliery manager and Alex MacDonald became district superintendent. The brattice was hung from the mine roof like curtains, and fresh air moved into the workings inside the brattice airway and returned from the faces on the other side of the fabric curtains.

The "pillar coal" was removed from the workings during the last century in Caledonia mines by "hand-pick-miners", and it became hazardous to mine the small pillars. The pressure busted the pillar and the coal face was lost, and the miners were refusing to work in the pillar sections. There was no machine mining in Caledonia during this time, but the circular mining, which was mined by a "Standely Header", could be noticed in "drifts" near the pit bottom, and a 500-pound puncher (a mining machine) was down in the east deep during the year 1900. The Standely mining machine was designed in England for underground development work by two men; it mined an annular grove in the coal seam, and it was operated by compressed air. The core of coal left after the header completed its circular "cut" was removed by explosives. The writer was told the Standely mining machine at Caledonia was the only one of its kind in Nova Scotia mines, and that machine was placed in a "cross-cut" near the pit bottom and never taken out of there.

Some of the lines written here are the actual notes of Alex MacDonald, and were written by him many years ago in the Caledonia mine office. Mr. MacDonald was the first Caledonia collier to be appointed a mine manager, and for many years, until the colliery was stopped, only young men from Caledonia were the underground officials and mine managers. That old mine manager has written the following: "This mine was opened in the present shaft and commenced mining operations during 1864 or 1865, by Mr. Poole, who was the mine manager and agent." The Phalen coal seam was worked before these dates at "Big Glace Bay" on land owned by a man named Phalen, from which the coal seam derived its name. Two shafts were sunk at the present site, and they both struck the Phalen at 165 feet from the surface. The coal seam was seven feet in height and dipped slightly from the south to a northerly direction. The seam was regular, only a few local faults, but 1,500 feet from the east side of the hoisting shaft an upthrow fault was met which pushed the coal seam seven feet above the regular angle of the coal now worked in this mine. This nonconformity decreased in the seam as the deep workings dipped towards the seacoast. There are no troublesome faults now in the Caledonia. The first coal mined and hoisted from this shaft was taken from the "rise", and hauled to the pit bottom by horses. Later a "jig wheel" was installed to shorten the horse haulage. A "jig" is a device which pulled mine cars by the acceleration of its own weight, and was in the early mines for many years. The most productive workings in the old Caledonia mine was worked on a similar system in the East Deep, and was known as "gravity section".

These lines are more notes from Mr. MacDonald's papers. One thousand feet from the coal hoist shaft the return air-shaft contained a bricked furnace that supplied the underground ventilation until the first mechanical fan was installed at this colliery during 1892. The mechanical ventilator produced 100,000 cubic feet of air a minute, and later a Dixon fan was placed in the intake air-shaft at Caledonia which pushed 150,000 cubic feet of air underground at a remarkable water gauge of only 3.5 inches. Such a low water gauge is a tribute to the underground officials who looked after the ventilating airways in Glace Bay's mines for many years. The maximum efficiency of any mine ventilator which is placed on the surface of underground may be lost through inefficiency of mine officials who never attended a colliery mining school or studied from a miner's textbook. It has been my endeavor throughout these pages to give the sound reasoning that mines can be worked for profits in safety by skilled colliers from the mining districts.

Mr. Stewart MacAulay's old scrapbook was in the Glace Bay Miners' Museum two years ago; it had a lot of information and some of it I used in this paper. Coal outputs were in the daily "Gazette" six days a week from all the collieries, but profits and prices taken from the tonnages were never published. Although it was well known, coal company stock certificates always paid the best dividends. The stories of hardship and amusement are not easy to come by in the mining districts, but there are old notes and a few diaries in trunks that have these anecdotes of the past. Most of the older miners are reluctant to talk about the "good old days". Once in a while an old collier may be persuaded to tell a few tales of the days gone by, and they laugh at themselves and others while describing things and people.

Standing on Pit Street at the top of old English town rows in Caledonia, an old collier related the following to me. "Do you remember when 'Donald Archie' returned from France soon after the first World War was over?" I did not, but I was listening. Well, he lived over there in the "Monkey Row" with his war bride. Donald was taking it easy for a few months on service pay, but he started to work and his name and number were on the Caledonia payroll. The story he was telling me happened in the "pluck-me" one Monday afternoon, and because Donald's wife was refused an order on Saturday morning, he would have been in there that afternoon but he was rabbit hunting that weekend. Donald Archy was shouting at the clerks in the grocery department when Mr. Shanks, the store manager, arrived on the scene and told the vexed miner that he had no "time" in the colliery office or any coal on the weighman's sheet, and he would not have another order until he was working every day the pit worked. Donald was a big man, over six feet tall — the store manager was only a few inches over five feet — and it was not much of an effort for the big man to over-power the smaller store manager. Donald Archy shook Mr. Shanks until his gold watch fell from his pocket, and the miner picked the watch off the floor and stated, "Mr. Shanks, you will get your watch when I get my order and it better be soon or this thing is going in the kitchen stove." What happened over this row? "Nothing." Donald sold a few rabbits and bought several items at Medjuck's store on Pit Street that evening, and later went to see the mine manager, Jack Casey, at his home and was working the next day. Mr. Shanks had his watch back in his pocket, but now secured with a gold chain.

There were stories that the colliers told each other in Gaelic for their own amusement that they would not tell outsiders, as many of these tales were "half truths and half lies". However, the story of "old Bony" was fresh in the minds of the Caledonia miners at the turn of this century. Bony was a Frenchman who arrived in Glace Bay from nowhere, and started to work in the Caledonia mine. It was not just gossip that the Frenchman had sold his soul to the devil for saving his life from a ship which was wrecked on the rocks at "Flint Island". Years ago I talked with an old collier who told me that he hauled coal from "Old Bony" when he was a young mine driver, and that he never went into the Frenchman's room, but his coal box was filled, yet he worked alone. Very few men have worked alone at the coal faces, maybe for a shift or two, but nobody would work with Bony, yet he was loading as much coal as two people. It was told that the miner talked with someone while he worked alone, and some were sure evil spirits were helping the man at his job; he continued on until a death bed repentance, but was buried outside the old Caledonia graveyard on Brookside Street. However, Satan did not give up easily. It was several years after "Bony" was buried that two young chaps, "on the dark of the moon", decided they were going to put the old man's remains inside the graveyard. They had just started to open the grave when a mad dog with "horse hooves"for paws chased them up to Caledonia Corner. From the stories that I have heard, "Old Bony" "was the greatest loader of them all", but people always reminded others "that the devil was good to his own."

HISTORIC PROPERTIES – GLACE BAY

"A" The "MONKEY ROW," at Caledonia, although the writer has known that many people call this building the "Holy Row".

"B" Bridgeport Rows - Some of these houses are now shown at the Miners Museum in Glace Bay. The tiny hamlet at the museum contains a company store, a fine dining room and an early coal miner's company house.

A

B

Before starting to write about the last century Caledonia colliers, there is an important matter that concerns the people who operate the Glace Bay Miners' Museum. I have inquired at Glace Bay about two machines which were operating at the mine when Caledonia closed. Why those machines? Because they are a part of the colliery's history, but I never received any information about them. During 1893 the World Fair was held in Chicago, Ill. The Rand Manufacturing Corporation showed a large "Air compressor" which took first prize in the heavy machinery exhibits. That compressor was purchased by the people who owned the Caledonia mine at the time. Then, in the year 1900, the Rand people sold another compressor which was shipped to Caledonia. What happened to both of these machines? Surely it is not possible that those two historic machines became filler for the old mine shaft; if not, why aren't they on exhibit at the Glace Bay Miners' Museum? "Another question." What happened to all the fossils which were placed in the "drager house" at Number 2 colliery in New Aberdeen?

The Caledonia colliery had two major fires between 1894 and 1899. The first was started during the enlargement of the coal-hoisting shaft when a lighted torch set coal dust afire. There was no loss of miners in the shaft fire, but several horses were smothered with smoke in the mine stables. The fire and the explosion during the month of June, 1899, was the worst disaster in the Glace Bay mining history as 11 colliers were killed in the mine. Like many other coal mine misfortunes where miners are fatalities, there is generally a mystery connected with the happenings, and it was told over the years that the underground manager was responsible for the blast, although he also died in the explosion. Nonetheless, no concrete mining evidence or history has stated this to be true. The fire started in the west deep pump house, and quickly spread into worked-out rooms which were developed from the west side of the mine. The 11 mine victims were in the pump house fighting a fire, and it is possible that they had succeeded in putting out the initial fire before they broke through a wooden stopping to follow the fire. It was later told in the mining districts that it was a mistake to break through that particular stopping. The explosion at Number 26 colliery on February 24, 1979, claimed 11 lives; 10 died in the mine and one in a Halifax hospital. Will it become another mystery in the colliery communities? The Number 26 pit is the old IB mine working on a new coal seam, the Harbor; it was a model colliery.

The Caledonia colliery became a model during the years it was worked by "room and pillar" method of mining, although it was never one of the big producers. The large producers were numbers IB and 2, at Glace Bay, and numbers 12 and 16 in the town of New Waterford, now all closed.

The early Caledonia miners leaving the pit bottom at five o'clock in the morning did not complain that they were loaded down for the long walk into the coal faces. These coal diggers had their "piece cans", you can be sure — a large canister full of tea, a can of water, a pocket full of chewing tobacco, approximately six pounds of explosives (bull-dog powder) in a buttoned-down case, and another receptacle which contained kerosene oil to refill their open flame lamps. The lamps were held in place on their foreheads with soft caps that miners wore underground before the days of "hard hats" in the coal pits. Going by the "pick house" these same people found two sharp picks, and a "breast auger" drill with their coal tally marked on them. Then they walked and talked for another mile into the mine workings. "And nowadays colliers are lammenting that they are 'weighted down" with an electric lamp and a gas mask at work, and a small package of gum in their pockets."

At one time at the Caledonia colliery, before strict laws were enforced on explosives in the mine and on the surface, qualified miners who held first-class mining certificates generally kept a weekly supply of "powder and squibs" in a porch off the kitchen at home, or in a wooden box in the outhouse. Written by hand, in ink, on the qualified miner's papers, was the following statement, "Competent Shot-lighter", and these people were allowed to "shoot down" their coal faces after it was mined and drilled. Not all the miners with "papers" had this declaration written on their certificates; that was obtained for them by the colliery agent or the mine manager.

Most of the mining and blasting was finished around ten o'clock, and the workers ate their breakfast away from the faces, as the smoke and dust was clearing from the "room-ends". After they had eaten their lunches, made a trip to an old "cross-cut", they started to load the coal that they mined. The coal was filled into small mine cars and hauled from the faces by horses that were driven by young lads from the community. In some mining districts these "children" were known as "mule skinners" or "pit boys". That system of mining was followed in Cape Breton mines for many years. However, when machines were introduced at the Caledonia colliery, the machine operators did not load the coal that they mined at the coal faces, but they were paid tonnage rates from the amount of coal their "loaders" sent to the pit-head. Several of the Caledonia "machine runners" during the last century were considered large wage earners who were paid monthly, and I have read an old pay script of 200 dollars. Not too bad for 100 years ago when

YOUNG COAL MINERS – CALEDONIA COLLIERY – 1903

datal rates were one dollar a day.

The Caledonia miners were open-handed with others in the mine workings during the early years; they loaned powder and tools to each other that they paid for on the colliery check-off at the end of the month. There were days that some colliers gave away most of their lunches, and it was a general custom in Cape Breton mines for the contract coal face workers to supply their driver with the oil he burned in his lamp and the tobacco he chewed underground.

The "tiny pit boy" shown in the picture of the young Caledonia miners, was a "mine driver", although he was born in Boston, Massachusetts, far from any coal mine shafts. Allie MacKenzie had a career in professional baseball. He was also a star rugby player and an officer in France during the first World War. Sometime after "Allie" returned from the war he became a mine official, but that little boy in the picture was fatally gassed in Number 24 mine many years later.

Horses trained for mine work were valuable property to coal companies during the many years which horse haulage was the means of transportation of minerals in the mines. However, horses, mules and ponies are used in mines today in many parts of the world. The mine "horse trainer" had the average horse ready to haul coal underground in one week, although it often took longer, and I have seen horses that worked satisfactorily after only three days in the mine darkness. Personally, I do not think the horses disliked the darkness too much, and it is possible that most of the smart mine animals used another sense in the coal pits. It is also well known to the mine horsemen that their horses sighted danger in the darkness before they did, and that this sense has saved the lives of drivers.

There were many stories around the collieries of the cruelty which the horses in the mines were subjected to; most of these tales were untrue, although I will admit it could have happened, and probably did, in some mines. Most of the underground drivers cared for their horses, and boasted to others what "Bob's" did, or could do in the darkness without a light. The old-time pit horse was never guided around the coal faces after he was unhooked from the mine car; he went to the coal face where he turned around and waited for his driver to hook him on to a full coal car. Only once in all the years that I have experienced with mine horses have I seen a horse severely whipped by a mine driver, and that lad was not very popular with the other young boys in Caledonia at that time. The drivers had whips that they made themselves, but pit horses were never beaten like race track horses.

"PIT HORSES"

The mine horses were taken to the surface during the colliers vacation periods. It was a habit at the collieries to repair the ventilating machinery during the miners holidays, and the horses would have a few days in the "pit yard", as the fans were stopped.

There were other factors which made the mine horses weary and tired underground. Broken roof timber scraped the horses' withers, but their backs and flanks were protected with leather back-pads. Harnesses not cleaned at regular intervals would cause sores on the animals' shoulders, and passing through narrow places scratched the skin off their sides. Like most people who learn from experience, the horses would not pass through close timber and bruise themselves several times and props had to be shifted to widen a safe passageway for them.

Although it was not a habit in Cape Breton mines to work horses "double shifts", they were, at times. The following lines were taken from an early Caledonia mine official; the fire in the "Hub" mine at New Aberdeen was the cause for this colliery to work on both shifts as most of the Hub miners were "taken on" here at Caledonia. There were 110 horses underground at that time and many of them worked the two shifts until new horses were purchased and trained for mine work. The coal pits in the Glace Bay district only worked on one production shift until the 1920s, and in the 1930s all the mines in Nova Scotia were producing coal on two shifts and have remained that way. However, the explosion which occurred at Number 26 mine during 1979 happened on the third production shift. For many years at Caledonia mines no explosives were allowed underground during the production shift, and the coal faces were "shot-down" on the afternoon maintenance shift. This was a safe practice, but several "shot-lighters" were working alone, and a fatality is noted in the Caledonia blasting records. Hugh MacDonald was struck by a small piece of coal and bled to death before he was found.

There were generally a few spare horses in the mine stables for production purposes, and one or two old horses which had light work on construction jobs. Horses are a lot like people, and become fat and lazy not working for long periods of time, and would have to be worked several days before they returned to former fitness. Some years ago in a Nova Scotia mine which used a large number of horses for "room and pillar" mining methods, an old horse was not counted in the annual "clean up" of non-working animals. That horse was hidden by the mine stableman, on the advice of an underground official, for over a year. Soon after the mine official retired, that old horse was taken to the surface and purchased by his old mine boss, who had a small farm in the Glace Bay district.

The next few lines were clipped from the Halifax **Mail Star**, during April 1978. "MINERS RESCUE UNDERGROUND BUDDY." A 12-year-old pit horse was left underground when the ventilation shaft collapsed at the old Drummond mine in

Pictou County. The miners escaped without injuries, but an old coal-black mare refused to follow the colliers to safety, and "Queenie" was left in the gassy darkness. Later, at great risks to themselves, several mine workers, accompanied by a veterinarian, went into the pit after Queenie. The mare was put asleep for a few hours and "tramed" to the surface. The mare, is now stabled above ground after nearly a decade in the mine darkness. Her eyesight was not injured, although some people think pit horses are blinded permanently by the sun after returning to the surface, with years working in the mine darkness.

The horses worked in the New Aberdeen mines were bigger than the animals used in Caledonia pit workings. The mine cars were larger and heavier in IB and Number 2 mines, then the Caledonia "pit coal box"; the latter approximately one and one-quarter tons, and the former over two tons. The pit horses were not controlled by reins, but they wore a leather protective bridle on their heads, and short reins of hemp rope which were tied to the back-pads. The horses listened and worked by the sound of their driver's voice in the semi-darkness, during the years that miners used the closed lights for illumination. The horse did whatever he was told, and he knew when his driver crossly checked him in a stern voice, but there were days the drivers were not talking, or the horses not listening which caused the derailment of full coal cars coming from the coal faces. It was stated that certain people did whip their horses after cars of coal left the tracks during "wild-runaways" on haulage roads after the animals were frightened by something. The mine driver went into the underground stables in the morning and put the bridle on his horse; the animal left the stall and the driver followed him into the mine workings. Seldom was a hand placed on him until harnessed into shafts for the rest of the shift. The shafts were two hardwood poles which were shaped like the shafts on race bikes, and bolted to an iron half circle bow. The mine horse shaft was an English mine carpenter's invention; before the shaft, hemp ropes were used which were tied to whiffletrees. The steel bow shafts were not used in Western Canada's mines during the underground horse haulage days, although the wooden shafts were safer than "traces and whiffletrees" for "mule-skinners".

Most of the horses which worked underground in the Nova Scotia mines were foaled on the Canadian Prairies, and many of them were shipped to the Eastern provinces and worked on farms for six months or a year before they were taken to the collieries. Wild horses were seldom taken underground and trained, but some "bad ones" became good working animals.

There were horses that did not like a change of drivers, and often "played on" young boys and had to be taken over by older drivers. However, horses with a "mean streak" in them were "sent up" after they kicked and bit other horses, and sometimes people. The average age for a horse when placed in the pits was five years, and the average animal lasted approximately seven years pulling mine cars from the coal faces. Most of the horses that worked in the "hoist" sections of Number IB mine were large horses, some as big as draft horses, and could not be used in the deeps below the motor road. There were no horses on the Western hemisphere before the Spaniards arrived in the New World. Nonetheless, horses were developed from a prehistoric animal, the Eohippus, which was native to America.

The horses that survived in the mines longer than the expected life of pit animals were generally sold to small farmers on the outskirts of the colliery districts. However, there was a fox farm operated by the O'Neils at Bridgeport at one time, but no old pit horse was intentionally butchered to feed foxes. Nonetheless, the foxes did eat a lot of horse meat at O'Neil's farm; animals which were severely injured in the pits were destroyed and taken to the fox farm. Last Summer the writer visited Marshfield, Mass., and read the following in a newspaper. The town's riding school horses were sold for dog meat and town's riding school was closed for the Summer. In Europe there are ponies in the coal pits, but in England they will soon all be removed from the mines. The British people are concerned about what is happening to the ponies after they are taken out of the coal mines, because it was stated that these animals were sold to Paris restaurants. Horse meat became a delicacy in Parisian rotisseries after the end of the second World War. That prompted a British member of Parliament to try for a "Bill" that would leave "pit ponies" which served their time underground in green pastures on the land owned by their government.

The mine horses that had accidents of a minor nature were kept in the underground stables for a few days, generally in a "box stall" with a foot of hay on the floor for bedding. The horses which had severe injuries in the Cape Breton mines were taken to the horse hospital at the Sterling Yard in Glace Bay, and cared for by the company horse doctor and his assistants. Most people today are fond of race horses, and they visit animal hospitals where skilled veterinarians are helping them, and they did, in years gone by, go to see the injured mine animals. However, at the mine hospitals people were told too often that the horses were intentionally injured in the pits. "Not so." The old pit horses in the Nova Scotia mines have received more loving care from their drivers than most surface horses, although they were worked hard in the mine darkness.

The Dominion Coal Company had at one time several hundred horses underground in clean stables, and they were fed and watered several times, before they entered and after returning from the mine workings. The best of hay, oats and water was available in their stalls in the Cape Breton mines. The horses working underground drink very little water, but they like to be hosed and washed with water in the stables, and I have seen drivers who had a "big day" do that before going to the surface. During the first years of mining, and for many decades later, the mine stables were close to the "pit bottoms", but as the mines advanced underground each section had its own stable. There were approximately 50 horses to be fed in each of the stables, and there were stablemen in there seven days a week and 24 hours a day.

It might seem odd today to hear the expression "horse sense", but many of the old mine horses had plenty of it in the underground darkness. It is a known fact that mine horses have refused to leave the stables and work the second shift, and I have read underground bosses' reports that stated a loss in production "because a horse refused to work."

The old British Mines Act states, "Every road where horses travel in the mines will be of sufficient width and height for these animals to pass in the underground working without rubbing or scraping themselves." The early mine horses coming from the Prairies in Western Canada were branded; later they were numbered on the inside of the lip. The sketch of measurements on the horses' hooves was placed below the animal's numbers in the surface blacksmith shops and the mine horseshoer made shoes to fit the numbers. Horses worked in the pits for several hours after losing a shoe, became lame, and could not be taken from the stables for a few days. The first underground stables in the Nova Scotia mines were not floored with wooden planks, but some were paved with concrete to level the floors and for drainage. Later most of the mine stables were made with wood floors, but very few of the horses laid down to rest underground.

I will conclude this little book with some observations on the functions and the fundamental purpose of the colliery mining schools in coal towns. Throughout these four articles I proclaimed the merits of the colliery mining classes. There is a significant difference between people in the mines who attended the colliery mining classes, and the mine foremen who have not done so. The general purpose for the mine schools in Nova Scotia is for their own miners' safety, but deep coal mining is going to advance all over North America and hundreds of learned colliers will be needed to guide others in the pits, in the near future. The Cape Breton colliers have gained a reputation in mining fields far from Glace Bay as safe miners and good workers, because their people advanced themselves through the colliery night schools and textbooks on coal mining. Mining laws must be supported by the people they protect, and the mine boss must enforce the mine act. Without these people, there will be no security in the mines.